Grade 4

Reveal MATH®

Student Practice Book

Mc
Graw
Hill

mheducation.com/prek-12

Send all inquiries to:
McGraw Hill
8787 Orion Place
Columbus, OH 43240

ISBN: 978-0-07-693709-7
MHID: 0-07-693709-7

Printed in the United States of America.

6 7 8 9 LHS 25 24 23 22

Grade 4
Table of Contents

Unit 6

Multiplication Strategies with Multi-Digit Numbers

Lessons

Unit 7

Division Strategies with Multi-Digit Dividends and 1-Digit Divisors

Lessons

Unit 8

Fraction Equivalence

Lessons

Unit 9

Addition and Subtraction Meanings and Strategies with Fractions

Unit 10

Addition and Subtraction Strategies with Mixed Numbers

Unit 11

Multiply Fractions by Whole Numbers

Unit 12

Decimal Fractions

Unit 13
Units of Measurement and Data

Lessons

Unit 14
Geometric Figures

Lessons

Additional Practice

Name _____

Review

You can round large numbers to any place.

Use a number line to round 384,264 to the nearest ten thousand.

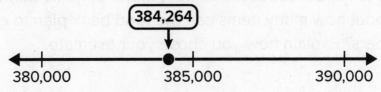

384,264 is between 380,000 and 390,000.

385,000 is halfway between 380,000 and 390,000.

384,264 is closer to 380,000.

384,264 rounded to the nearest ten thousand is 380,000.

---

1. Round the following numbers to the nearest hundred thousand, ten thousand, and thousand.

	hundred thousand	ten thousand	thousand
125,250			
485,649			
518,341			
826,341			

2. A company is mailing out 16,743 letters. Envelopes are sold in boxes of one thousand. How many envelopes will the company purchase in order to have enough for all of the letters? Explain.

3. A food bank collects 324,887 items in one year. The food bank expects to collect about the same number of items during the next year. About how many items can the food bank plan to collect in the 2 years? Explain how you chose your estimate.

4. A school district gives each student an agenda book at the beginning of the school year. There are 163,641 students currently enrolled in the district. Estimate the number of agenda books the district should buy in order to account for any new students that arrive throughout the year. Explain your estimate.

Math @ Home Activity

Identify large numbers in magazines or online articles. Ask your child to round each number to the nearest hundred thousand, ten thousand, and thousand. Then ask him or her to identify which estimate would be the most acceptable in each situation.

Additional Practice

Name _____

Review

You can use estimates to decide whether an answer is reasonable. You can estimate a sum or difference by rounding numbers, using compatible numbers, or thinking about a range.

A park ranger records the sightings of different animals at the park. He sees 347 reptiles and 1,191 mammals within one week. About how many more mammals than reptiles did he see?

Round to estimate. $1,190 - 350 = 840$

Use compatible numbers to estimate. $1,200 - 350 = 850$

Think about a range.   Between 700 and 900.

---

1. Estimate. Then use a calculator to find the actual sum or difference.

	Estimate	Actual Answer
$749 + 2,810 = ?$		
$4,968 - 354 = ?$		
$3,583 + 5,685 = ?$		
$14,632 - 10,766 = ?$		

2. A shipping company ships 934 small packages and 4,634 large packages. About how many packages does the company ship altogether?

3. A grocery store sells 1,163 bananas and 2,891 apples in a week. About how many more apples than bananas does the grocery store sell?

4. A banker estimates the difference of $3,417 − $1,092 to be $2,330. Do you agree or disagree with the estimate? Explain.

5. Estimate the range that the sum of 2,195 and 985 will be in. Show your work.

6. Arnold estimates 8,438 − 5,534 to be 2,910. Luetta estimates the difference to be 2,900. Can both estimates be correct? Explain.

Help your child find a table of values in a magazine or book. The numbers in the table should be three to five digits long. Create word problems for your child to solve that involve using the numbers in the table to estimate sums and differences. Allow him or her to use a calculator, then determine if their calculated sum is reasonable.

Additional Practice

Name _____

Review

You can add multi-digit numbers in different ways.

Add 1,526 + 1,294.

One way to add is using partial sums.

Add the ones	6 + 4	10
Add the tens	20 + 90	110
Add the hundreds	500 + 200	700
Add the thousands	1,000 + 1,000	2,000
Add the partial sums.		2,820

1,526 + 1,294 = 2,820

Another way is to make adjustments.

1,526 + 1,294

+4   −4

1,530 + 1,290 = 2,820

Solve and show your work.

1. 547 + 3,213 = _____

2. 11,372 + 6,284 = _____

3. 2,865 + 1,925 = _____

4. 375 + 6,242 = _____

5. 124,326 + 34,574 = _____

6. An arcade had 1,933 visitors in March and 1,506 visitors in April. What is the total number of people that visited the arcade in March and April?

7. A school held a fundraiser and raised $11,214 last year. This year they raised $12,587. What is the total amount raised over these two years?

8. The local zoo has two elephants, one male and one female. The male elephant weights 15,117 pounds. The female elephant weighs 13,945 pounds. What is the combined weight of the elephants?

Math @ Home Activity

Give your child 4 crayons, markers, or pencils that are different colors. Ask him or her to find the sum of 31,426 + 18,445 by using partial sums. He or she should color code the partial sums by place value so it is easy to see the values. Repeat the activity with different addition problems as long as time permits.

Additional Practice

Name _____

Review

You can use an algorithm, or a set of steps, to add numbers.

Find 3,451 + 1,317.

$$
\begin{array}{r}
3{,}451 \\
+\,1{,}317 \\
\hline
4{,}768
\end{array}
$$

Add the ones digits first. $1 + 7 = 8$

Then add the tens digits. $5 + 1 = 6$

Add the hundreds next. $4 + 3 = 7$

Add the thousands last. $3 + 1 = 4$

Solve. Show your work.

1. $\begin{array}{r} 2{,}365 \\ +\ \ \ 423 \\ \hline \end{array}$

2. $\begin{array}{r} 5{,}617 \\ +\ \ \ 372 \\ \hline \end{array}$

3. $\begin{array}{r} 7{,}408 \\ +\,1{,}211 \\ \hline \end{array}$

4. $\begin{array}{r} 8{,}371 \\ +\,1{,}522 \\ \hline \end{array}$

5. $\begin{array}{r} 21{,}396 \\ +\ \ 5{,}401 \\ \hline \end{array}$

6. $\begin{array}{r} 33{,}544 \\ +\ \ 2{,}135 \\ \hline \end{array}$

7. $\begin{array}{r} 26{,}496 \\ +\,11{,}502 \\ \hline \end{array}$

8. $\begin{array}{r} 46{,}837 \\ +\,22{,}141 \\ \hline \end{array}$

9. A library donates 1,124 new books and 571 movies to a school. How many items does the library donate?

10. Olivia has 2,130 songs on her computer and 742 songs on her phone. How many songs does she have in all?

11. A grocery store makes $12,571 in January. In February, the store makes $14,305. How much money does the store make in January and February?

12. Natalya walks 13,421 steps on Monday and 15,364 steps on Tuesday. Otelia walks 13,537 steps on Monday and 15,412 steps on Tuesday. Who walks the greater number of steps? Explain.

Ask your child to state the order in which the digits in a multi-digit addition problem are added. Use everyday situations to create addition problems that require no regrouping for your child to practice using the algorithm. For example, if you keep track of the number of calories consumed in a day, have your child find the calorie count for 2 meals.

Additional Practice

Name _____

Review

You can add large numbers with an algorithm. You write the equation vertically and add the digits in the same place value. When the sum of the digits in one place is a 2-digit number, you need to regroup.

A small company produces 12,458 pencils and 6,235 pens in one day. How many pencils and pens does the company produce in one day?

$$
\begin{array}{r}
\overset{1}{1}2{,}458 \\
+\ \ 6{,}235 \\
\hline
18{,}693
\end{array}
$$

Add the ones. The sum of 8 ones and 5 ones is 13 ones. You record the 3 of the 13 ones and regroup the ten.

Add the tens. No regrouping is required.

Add the hundreds. No regrouping is required.

Add the thousands. No regrouping is required.

Add the ten thousands.

What is the sum? Show your work.

1.
$$
\begin{array}{r}
1{,}642 \\
+\ \ \ 387 \\
\hline
\end{array}
$$

2.
$$
\begin{array}{r}
2{,}094 \\
+\ 1{,}037 \\
\hline
\end{array}
$$

3.
$$
\begin{array}{r}
5{,}947 \\
+4{,}506 \\
\hline
\end{array}
$$

4.
$$
\begin{array}{r}
18{,}761 \\
+2{,}390 \\
\hline
\end{array}
$$

5.   13,483
       + 14,918

6.   19,798
       + 10,403

7.   26,917
       + 17,015

8.   20,685
       + 18,832

9.   29,731
       + 25,933

10.   28,675
        + 24,739

---

11. Alene scores 2,467 points during round 1 of a board game and 3,946 points during round 2. How many points does she score in all?

12. Delma's work is shown below. How do you respond to her calculations? Explain.

   ^{1 1}
  13,687
+ 3,595
  17,272

Math @ Home Activity

Have your child roll a number cube 10 times, recording the numbers in the following format.

xx,xxx
+ xx,xxx

Have him or her add the numbers using the algorithm for addition. Then check your child's work. Have your child repeat this activity a few times.

Additional Practice

Name _____

Review

You can subtract multi-digit numbers by decomposing the number you are subtracting and then subtract by place value.

$7,855 - 3,420 = ?$

Decompose the number: $3,420 = 3,000 + 400 + 20$

Subtract: $7,855 - 3,000 = 4,855$

$4,855 - 400 = 4,455$

$4,455 - 20 = 4,435$

$7,855 - 3,420 = 4,435$

Solve and show your work.

1. $2,985 - 1,270 =$ _____

2. $7,968 - 4,126 =$ _____

3. $5,438 - 3,215 =$ _____

4. $67,855 - 33,420 =$ _____

5. A school bus driver drives 2,176 miles in December and 3,241 miles in January. How many more miles does the bus driver drive in January compared to December?

6. Caldon School District has 12,064 students enrolled. There are 10,933 students enrolled in Dempsey School District. How many more students are enrolled in Caldon School District than in Dempsey School District?

7. A school sold 5,618 tickets to the first football game of the season. There were 4,724 tickets sold to the second football game. How many more tickets were sold for the first football game?

8. There were two students running for student body president. The first student received 3,217 votes. The second student received 2,987 votes. How many more votes did the first candidate receive?

Math @ Home Activity

Using the internet or a newspaper, find the prices of different cars, trucks, or SUVs. Have your child find the difference between the prices of 2 vehicles using decomposing or adjustments. Have him or her explain how they found the difference.

Additional Practice

Name _____

Review

You can use an algorithm to subtract multi-digit numbers. First, write the equation vertically and then subtract like units in each place-value position, starting with the ones place.

A theme park sells 4,572 adult tickets and 2,350 child tickets in one week. How many more adult tickets than child tickets sold?

$$\begin{array}{r} 4{,}572 \\ -\ 2{,}350 \\ \hline 2{,}222 \end{array}$$

Subtract the ones digits.

Subtract the tens digits.

Subtract the hundreds digits.

Subtract the thousands digits.

So, 2,222 more adult tickets were sold.

Solve using an algorithm.

1.
$$\begin{array}{r} 1{,}961 \\ -\ \ \ 510 \\ \hline \end{array}$$

2.
$$\begin{array}{r} 3{,}598 \\ -\ 2{,}485 \\ \hline \end{array}$$

3.
$$\begin{array}{r} 5{,}733 \\ -\ 3{,}213 \\ \hline \end{array}$$

4.
$$\begin{array}{r} 8{,}649 \\ -\ 5{,}407 \\ \hline \end{array}$$

5.
$$\begin{array}{r} 19{,}789 \\ -\ \ 5{,}364 \\ \hline \end{array}$$

6.
$$\begin{array}{r} 24{,}087 \\ -\ 13{,}055 \\ \hline \end{array}$$

7.   31,875
    − 11,744

8.   42,769
    − 21,254

9.   55,980
    − 34,560

10.   61,936
    − 21,721

---

11. Renetta buys a laptop that costs $1,230. She has $2,350 to spend. How much money does she have left?

12. Ariana walks 5,798 feet and runs 3,634 feet. How many more feet does she walk?

13. Company A makes $12,560 a month. After donating some money to an animal shelter, they have $11,430 left. Company B makes $13,679 a month. After donating some money to a food pantry, they have $12,552 left. Which company donates more money? Explain your reasoning.

Math @ Home Activity

Provide opportunities for your child to subtract using the algorithm by creating word problems based on everyday situations. These subtraction problems should *not* require regrouping. While your child works on one problem, work on another problem and make a mistake. Ask your child to identify and fix the mistake.

Additional Practice

Name _____

Review

You can use an algorithm for subtracting multi-digit numbers with regrouping. First write the equation vertically and then subtract the digits in the same place value.

Find 56,734 − 28,921.

$$
\begin{array}{r}
\overset{\scriptstyle 15}{} \\
^{4}\overset{5}{}\,^{17} \\
56,734 \\
-\ 28,921 \\
\hline
27,813
\end{array}
$$

Subtract the ones digits. Subtract the tens digits.

Regroup to subtract the hundreds digits.

Regroup to subtract the thousands digits.

Subtract the ten thousands digits.

Solve using an algorithm.

1.
$$
\begin{array}{r}
2,461 \\
-\ \ \ 397 \\
\hline
\end{array}
$$

2.
$$
\begin{array}{r}
3,309 \\
-\ 1,857 \\
\hline
\end{array}
$$

3.
$$
\begin{array}{r}
7,910 \\
-\ 4,761 \\
\hline
\end{array}
$$

4.
$$
\begin{array}{r}
18,847 \\
-\ \ 6,995 \\
\hline
\end{array}
$$

5.
$$
\begin{array}{r}
20,955 \\
-\ 18,293 \\
\hline
\end{array}
$$

6.
$$
\begin{array}{r}
26,008 \\
-\ 14,350 \\
\hline
\end{array}
$$

7. 31,701
   − 19,568

8. 43,998
   − 25,479

9. 61,597
   − 24,600

10. 80,645
    − 32,470

---

11. Sonny has 2,265 trading cards. His organizer holds 1,575 cards. How many cards will *not* fit in the organizer?

12. Florence has 7,615 black and white photos and 3,978 color photos. How many more black and white photos does she have?

13. Setsuko's work is shown at the right. What did he do wrong? What is the correct answer?

54,392
− 38,576
24,224

Copyright © McGraw-Hill Education

Additional Practice

Name _____

Review

You can use a bar diagram and an equation with a variable to represent and solve a multi-step problem.

The zoo had students visiting during the first three days in May. On the first day 2,915 students visited. On the second day 412 fewer students visited. Attendance records showed 175 fewer students visited on the third day than the second day. How many total students visited over those three days?

Step 1:

First day

2,915

$s = 2,915 - 412$

Second day

s	--- 412 ---

$s = 2,503$

Step 2:

Second day

2,503

$t = 2,503 - 175$

Third day

175	t

$t = 2,328$

Step 3:

a

2,915	2,503	2,328

$a = 2,915 + 2,503 + 2,328$

$a = 7,746$

Use diagrams and equations to solve each problem. Show your work.

1. A company printed flyers to advertise their upcoming sale. The flyers were printed in 3 batches. The first batch had 7,355 flyers. The second batch had 1,690 fewer flyers than the first batch. The third batch had 895 fewer than the second batch. How many flyers did they print in all?

2. There are 625 students going on a field trip. The students are into four groups. How many students are in the red group?

Group Name	Number of students
Blue	144
Green	162
Yellow	155
Red	?

3. A food bank collected 3,887 food items in one month. The next month they collected 997 more food items. The following month they collected 572 fewer items than the first month. How many food items were collected over three months?

4. The water park had 15,276 visitors on opening weekend. The following weekend had 1,231 fewer visitors than opening weekend. The third weekend had 765 more visitors than the second weekend. How many visitors did the park have over the first three weekends?

Math @ Home Activity

Identify greater numbers around your home or in magazines and online articles. Ask your child to round the number to the nearest hundred thousand, ten thousand, and thousand. Then ask your child to identify which estimate is most reasonable in each situation.

Additional Practice

Name _____

Review

You can solve problems using several steps and multiple strategies.

Lakeside School District has $155,837 set aside to buy new playground equipment. They spent $87,230 on swing sets at all of the elementary schools in the district. They need $62,561 for slides and $30,652 for jungle gyms. How much more money do they need to purchase the slides and jungle gyms?

Step 1 Find the amount the school district has left after buying swing sets for all of the elementary schools.

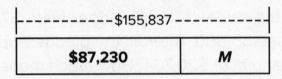

$M = 155,837 - 87,230$
$M = 68,607$

The district has $68,607 left.

Step 2 Find the cost of the slides and jungle gyms.

$62,561 + 30,652 = 93,213$

The slides and jungle gyms will cost $93,213.

Step 3 Find how much more money the group needs.

$93,213 - 68,607 = 24,606$

The district needs $24,606 more for new slides and jungle gyms.

Solve. Show your work.

1. The Green Meadows Homeowners Association had $4,715 in their account at the end of last year. This year they want to landscape the neighborhood park. In January they collect $1,348 in dues. The new landscaping costs $5,215. Will they have enough money for this project?

2. The New City Orchestra sold 2,736 tickets in advance for opening weekend. On Friday, they sold 3,835 tickets. They sold a total of 10,922 tickets. How many tickets did they sell on Saturday?

3. Brookfield Elementary School is holding a fun run. Each student has three weeks to get pledges to raise money for charity. After the first week the students' total pledges were $7,462. On the final week they raised $5,890. After all the pledges were collected the school raised a total of $20,704. How much money did they raise during the second week?

4. Kayla and her family are taking a four-day train trip across the country. They will be traveling a total of 2,787 miles. On the first and second day they traveled 825 miles total. On the third day they traveled 1,043 miles. How many miles left do they have to travel on the fourth day?

Math @ Home Activity

Tell your child you are going to plan a trip and have a budget of $300 a person for a plane ticket. Look online to find the cost of plane tickets to your destination. Determine if you have enough budgeted for everyone's plane tickets.

Additional Practice

Name _____

Review

You can write a multiplicative comparison statement that compares two or more quantities.

Stick A:

Stick B:

Multiplicative comparison statement: 12 cubes are 2 times as many as 6 cubes.

You can write a multiplication equation to represent the multiplicative comparison statement.

Equation: $2 \times 6 = 12$

---

1. Which comparison statement represent the equation $4 \times 8 = 32$? Choose all that apply.

   A. 4 is 8 times as many as 32.

   B. 32 is 4 times as many as 8.

   C. 4 is 4 times as many as 32.

   D. 8 is 8 times as many as 32.

2. Which of the following statements is true?

   A. 24 is 4 times as many as 6.

   B. 4 is 6 times as many as 24.

   C. 6 is 4 times as many as 24.

   D. 4 is 24 times as many as 6.

3. Write a multiplicative comparison statement to describe this comparison.

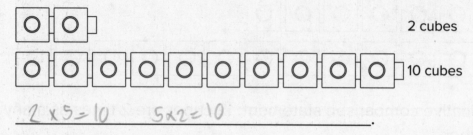

3 cubes

12 cubes

_____3_____ is _____4_____ times as many as _____12_____.

4. Write a multiplication equation to represent this comparison.

2 cubes

10 cubes

_____2 x 5 = 10   5 x 2 = 10_____.

5. Write a multiplication equation that matches this multiplicative comparison statement:

63 is 7 times as much as 9.

_____7 x 9 = 63_____.

6. Macy makes three sticks with connecting cubes. Stick A has 3 connecting cubes. Stick B has 6 connecting cubes. Stick C has 15 connecting cubes. Write multiplicative comparison statements to describe how Stick A compares to the other two sticks.

Math @ Home Activity

Gather some small objects that can attach to each other, such as paper clips. Attach the objects so there is one row of 2 objects and a second row of 8 objects. Then have your child write a multiplicative comparison statement and a multiplication equation to compare the two quantities.

Additional Practice

Name _____

Review

You can compare two quantities by telling how many more or how many times as many.

Nick hikes 2 miles. Logan hikes 4 more miles than Nick. Ava hikes 3 times as many miles as Nick. How many miles do Logan and Ava hike?

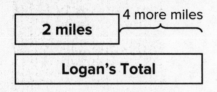

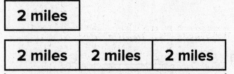

$2 + 4 =$ Logan's miles $2 \times 3 =$ Ava's miles

$2 + 4 = 6$ $2 \times 3 = 6$

Logan hikes 6 miles.   Ava hikes 6 miles.

Choose the correct comparison statement.

1. A small flagpole is 15 feet tall. A taller flagpole is 4 times as tall as the small flagpole. How tall is the taller flagpole?

    $15 + 4 = f$ $15 \times 4 = f$

2. Avery's green lizard eats 5 crickets. Her brown lizard eats 6 more crickets than her green lizard. How many crickets does her brown lizard eat?

    $5 + 6 = c$ $5 \times 6 = c$

Write equations to represent the comparison. Solve.

3. Twyla picks 8 flowers. Natalie picks 4 times as many flowers as Twyla. Ava picks 12 more than Twyla. How many flowers do Natalie and Ava pick?

Natalie picks _____ flowers. Ava picks _____ flowers.

4. A flamingo at a zoo is 4 feet tall. A giraffe at the zoo is 5 times as tall. How tall is the giraffe?

The giraffe is _____ feet tall.

5. Caleb reads 7 pages a day. Nieves reads 3 times as many pages as Celeb in a day. How many pages does Nieves read each day?

Nieves reads _____ pages each day.

6. Tia has 9 math problems for homework. Her brother has 3 times as many math problems for homework. Her sister has 6 more math problems for homework. How many math problems do Tia's brother and sister have for homework?

Have your child practice solving comparison word problems using addition or multiplication. For example, have your child use dry pasta to answer problems like the following.

Kim has 6 CDs. Ty has 4 times as many CDs as Kim. How many CDs does Ty have?

Lesson **4-3**

Additional Practice

Name _____

Review

You can represent multiplicative comparison problems using bar diagrams and equations.

Owen serves 7 aces during a volleyball tournament. Emma serves 3 times as many aces as Owen. How many aces does Emma serve?

You can represent the problem with a bar diagram.

Owen | **7 aces** |

Emma | **7 aces** | **7 aces** | **7 aces** |

You can write a multiplication equation to represent the bar diagram.

$7 \times 3 = a$

So, Emma serves 21 aces during the volleyball tournament.

---

1. There are 3 cars in a parking lot. There are 5 times as many cars in a parking garage. How many cars are in the parking garage?

   Choose the equation that represents the problem.

   A. $5 \times b = 3$ **C.** $3 + 5 = c$

   B. $3 \times b = 5$ **D.** $3 \times 5 = c$

   How many cars are in the parking garage? _____

Draw a representation and write an equation to find the unknown. Use a symbol for the unknown. Solve.

2. Marcy studied 4 hours last week. She studies 4 times as many hours this week. How many hours does Marcy study this week?

   Marcy studies _____ hours this week.

3. One piece of fabric is 8 feet long. A second piece of fabric is 4 times as long. How long is the second piece of fabric?

   The second piece of fabric is _____ feet long.

4. Jackie has 9 stamps. Kevin has 36 stamps. Kevin says he has 5 times as many stamps as Jackie. Is Kevin correct? Explain your reasoning.

5. Erin picked 8 flowers. Jason picked 5 times as many flowers as Erin. How many flowers did Jason pick? Explain.

 **Math @ Home Activity**

Look for common household objects your child can use to practice solving multiplicative comparison problems. For example, have your child determine how many pens you have if he or she has 3 pens and you have 4 times as many pens.

Additional Practice

Name _____

Review

You can represent the multiplicative comparison with a bar diagram and a division equation.

A sporting goods store sells soccer cleats for $21. Soccer cleats cost 3 times as much as a pair of shin guards.

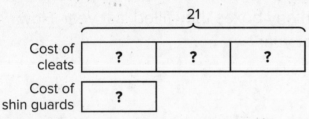

$21 \div 3 = ?$ $21 = 3 \times ?$

$21 \div 3 = 7$

So, the shin guards cost $7

Choose the division equation that represents each comparison.

1. 36 is 9 times as much as ?

   A. $36 \div 9 = 6$

   B. $36 \div 9 = 3$

   C. $36 \div 9 = 4$

   D. $36 \div 9 = 9$

2. 48 feet is 6 times as long as ?

   A. $48 \div 6 = 8$

   B. $48 \div 6 = 6$

   C. $48 \div 6 = 12$

   D. $48 \div 6 = 4$

Draw a bar diagram and write a division equation to represent each comparison.

3. Peta cuts a string 32 inches to make a necklace. A necklace is 4 times longer than a bracelet. How long is a bracelet?

4. A school food drive collects enough food to fill 27 boxes to donate. This is 3 times as many boxes they filled last year. How many boxes did they fill last year?

Write a division equation to represent the comparison. Solve.

5. Kai'noa uses 6 cups of flour to make bread. He uses 2 cups of sugar. How many times more flour does he use than sugar?

6. Suzanna bakes 45 blueberry muffins for the bake sale by increasing the number of ingredients need for the recipe. This is 5 times the number of muffins that the original recipe makes. How many muffins does the original recipe make?

Math @ Home Activity

Look at recipes with your child. Have the students write equations or draw bar diagrams to show how to compare different quantities. For example, if a recipe makes 24 granola bars, how many times more is that than a recipe that makes 8 granola bars?

Additional Practice

Name _____

Review

You can use an array to represent decomposing a number into factors.

Laquanda has 15 stamps she wants to arrange in a picture frame. She wants each row to have the same number of stamps. How can Laquanda arrange the stamps?

   **15 = 1 × 15**

Laquanda can arrange **15** stamps in **1** row of **15** or **15** rows of **1**.

   **15 = 3 × 5**

Laquanda can arrange **15** stamps in **3** rows of **5** or **5** rows of **3**.

1. Write factor pairs for each product.

    a. 8 _____

    b. 14 _____

    c. 24 _____

    d. 45 _____

    e. 50 _____

    f. 55 _____

    g. 65 _____

    h. 71 _____

Use what you know about multiplication and division to solve.

2. Angela is arranging 32 pictures in a photo album. She wants to have the same number of pictures in each row. How can she arrange her pictures?

3. Milo is displaying 20 framed photos on shelves. He wants the same number of photos on each shelf. He has up to 4 shelves to use. How can he arrange the photos?

4. Jenine is stacking boxes in a warehouse. She says she can arrange 17 boxes in equal rows of 4. Is Jenine correct? Explain your reasoning.

Math @ Home Activity

Have your child use common household objects to practice finding the factors of a given number. For example, give your child 12 books to arrange on up to 4 shelves. Tell your child to find as many ways to arrange the books as possible so the same number of books is on each shelf. Then have your child identify all of the factors of 12.

Additional Practice

Name _____

Review

You can classify a number as either composite or prime based on the number of factor pairs it has.

A school is arranging 9 soccer and 7 basketball trophies in a trophy case.

The soccer trophies can be arranged in 3 different ways.

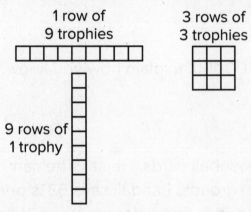

1 row of
9 trophies

3 rows of
3 trophies

9 rows of
1 trophy

The basketball trophies can be arranged in two ways.

1 row of
7 trophies

7 rows of
1 trophy

Determine if the number is <u>prime</u> or <u>composite</u>. Explain your reasoning.

1. 35 _____

2. 73 _____

3. 39 _____

4. 51 _____

Tell whether the statement is true or false. Then justify the answer.

5. Odd numbers cannot be composite.

6. All even numbers are composite.

7. Find a prime number between 80-90. Explain how you know it is prime.

8. Find a prime number between 100-150. Explain how you know it is prime.

9. Kevin wants to arrange his 63 baseball cards. He says he can arrange them in 6 different even groups. Randall says 63 is prime and the cards can only be arranged in two different group. Who is correct? Explain why.

10. Jasmine has 41 beads to display. She thinks that they can be arranged in more than 2 equal groups Is she correct? Explain.

Math @ Home Activity

Give your child a handle of beads or dry pasta. Ask them to count them and then determine how many different ways they can arrange them in equal rows. Then ask them to explain if there is a composite or prime number of items.

Additional Practice

Name _____

Review

A multiple of a number is the product of that number and any whole number. You can use equations to find the multiples of a number.

Find the first ten multiples of 4.

$1 \times 4 = 4$	$5 \times 4 = 20$	$9 \times 4 = 36$
$2 \times 4 = 8$	$6 \times 4 = 24$	$10 \times 4 = 40$
$3 \times 4 = 12$	$7 \times 4 = 28$	
$4 \times 4 = 16$	$8 \times 4 = 32$	

4, 8, 12, 16, 20, 24, 28, 32, 36, and 40 are multiples of 4.

Write the multiples in the blanks.

1. Write the first seven multiples of 2.

2, _____, _____, _____, _____, _____, _____

2. Write the first seven multiples of 5.

5, _____, _____, _____, _____, _____, _____

3. Write the first seven multiples of 6.

6, _____, _____, _____, _____, _____, _____

4. Write the first seven multiples of 8.

8, _____, _____, _____, _____, _____, _____

5. Which number is *not* a multiple of 7? Choose the correct answer

    A. 14   **C.** 49

    B. 17   **D.** 70

6. Which numbers are multiples of 9? Choose all that are.

    A. 90   **C.** 29

    B. 27   **D.** 72

Use what you know about multiples to solve.

7. Kiara buys 5 packs of trading cards to give to her friends. Each pack contains 8 cards. How many cards does Kiara buy?

8. An art teacher is handing out paintbrushes to students. If 4 students sit at each table, how many paintbrushes will she need for 3 tables of students? 4 tables of students? 5 tables of students?

9. Dominica is making gift bags. Each gift bag will have 3 items. How many items does she need if she wants to make 3 gift bags? 5 gift bags? 7 gift bags? Explain.

Math @ Home Activity

Identify products that are sold in multiples at your local grocery store. Have your child determine how many total products would be purchased if 1-6 packages were bought. For example, if yogurt cups are sold in packages of 6, ask your child how many yogurt cups would be purchased if you bought 1-6 packages.

Additional Practice

Name _____

Review

You can create a pattern by using a sequence of shapes or numbers that repeats the same process over and over again. A pattern can repeat or grow. Every pattern has a pattern rule that describes the pattern.

Vivianna creates the following pattern.

The pattern rule that Vivianna is following is square, circle, star, circle.

Complete the next three numbers in each sequence. Then explain the rule.

1. 16, 24, 32, 40, …

2. 3, 6, 12, 24, …

3. 49, 42, 35, 28, …

4. 90, 75, 60, 45, …

Describe the rule for each pattern.

5.

6. 22, 34, 46, 58, ...

7.

8. 5, 15, 45, 135

Extend each sequence by three terms. Explain how you found your answer.

9. 26, 39, 52, _____, _____, _____

10.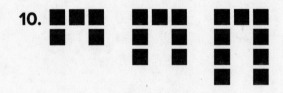

Create various patterns with your child that involve either numbers or shapes. If you create the pattern, have your child identify it. If your child creates the pattern, you identify it. When creating a number pattern, use addition, subtraction, multiplication, or division.

Additional Practice

Name _____

Review

You can analyze a pattern to find features that are not stated in the pattern rule.

Will the 15th circle be solid or have a pattern?

One Feature	Another Feature
Every other circle has a pattern on it.	All the odd circles are solid colored.

The 15th term will be solid because it is an odd number.

Select true or false about the pattern.

1. Start with 5, multiply by 3.

	True	False
All terms are multiples of 5		
All terms are multiples of 3		
All terms are even numbers		

2.

	True	False
Each row grows by 1		
The circles always add up to an odd number		
The rows alternate from odd to even		

Use the pattern to answer questions 3–4

24, 36, 48, 60, 72, ...

3. What is the pattern rule?

4. What is a feature of that pattern that is not stated in the pattern rule? Explain why this feature exists.

5. Zoe creates a square with 3-inch sides. She increases the length of each side by 2 inches to make each new square. If she continues the pattern, what will be the area of the 4th square?

Use items around the house to create a pattern. Ask your child to identify the pattern rule and then find two other features of that pattern not stated in the rule. After he or she finished create another pattern and repeat the activity.

Additional Practice

Name _____

Review

You can use arrays to show an equation and represent decomposing a factor.

When finding 5 × 7, you can decompose 7 in different ways.

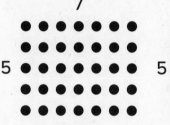

$5 \times 7 = \mathbf{35}$

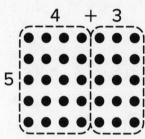

$5 \times 7 = 5 \times 4 + 5 \times 3$
$= 20 + 15$
$5 \times 7 = \mathbf{35}$

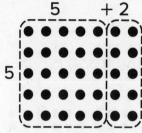

$5 \times 7 = 5 \times 5 + 5 \times 2$
$= 25 + 10$
$5 \times 7 = \mathbf{35}$

Decomposing 7 helps you find that 5 × 7 = 35.

Choose all the correct expressions that show how to decompose the factor.

1. 5 × 9

    A. (5 × 5) + (5 × 4)   **B.** (5 × 7) + (5 × 2)

    C. (5 × 3) + (5 × 2)   **D.** (5 × 8) + (5 × 1)

2. 4 × 7

    A. (4 × 2) + (4 × 2)   **B.** (4 × 4) + (4 × 3)

    C. (4 × 5) + (4 × 2)   **D.** (4 × 7) + (4 × 7)

Draw circles to divide the array into two groups to decompose a factor. Then complete the equations.

3. Find 6 × 8 by decomposing 8.

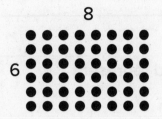

8

6

$6 \times 8 = (6 \times \underline{\hspace{1cm}}) + (6 \times \underline{\hspace{1cm}})$

$6 \times 8 = \underline{\hspace{1cm}}$

4. Find 8 × 9 by decomposing 9.

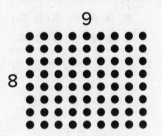

9

8

$8 \times 9 = (8 \times \underline{\hspace{1cm}}) + (8 \times \underline{\hspace{1cm}})$

$8 \times 9 = \underline{\hspace{1cm}}$

5. Find 9 × 9 by decomposing one of the factors.

$9 \times 9 = (\underline{\hspace{1cm}} \times \underline{\hspace{1cm}}) + (\underline{\hspace{1cm}} \times \underline{\hspace{1cm}})$

$9 \times 9 = \underline{\hspace{1cm}}$

6. Carson thinks you can find 4 × 9 by using (4 × 4) + (4 × 4). How would you respond to Carson?

Math @ Home Activity

Have your child use small objects, such as pennies, to create an array to represent a multiplication problem. Then tell your child to separate the pennies to represent decomposing one of the factors. Ask your child to identify the product.

Additional Practice

Name _____

Review

You can use the Distributive Property of Multiplication to decompose factors and find partial products.

Find 7 × 23.

$m = 7 \times 23$

Decompose a factor to make compatible numbers.

$m = 7 \times (20 + 3)$

$m = (7 \times 20) + (7 \times 3)$

Add partial products to find the product.

$m = 140 + 21$

$m = 161$

Use partial products to complete the equations.

1. $6 \times 18 = (6 \times \underline{\hspace{1cm}}) + (6 \times \underline{\hspace{1cm}})$

   $6 \times \mathbf{18} = \underline{\hspace{1cm}} + \underline{\hspace{1cm}}$

   $6 \times \mathbf{18} = \underline{\hspace{1cm}}$

2. $5 \times 26 = (5 \times \underline{\hspace{1cm}}) + (5 \times \underline{\hspace{1cm}})$

   $5 \times \mathbf{26} = \underline{\hspace{1cm}} + \underline{\hspace{1cm}}$

   $5 \times \mathbf{26} = \underline{\hspace{1cm}}$

Complete the models and equations.

3. 8 × 19

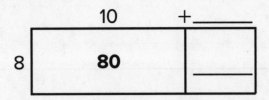

$8 \times \textbf{19} = (8 \times \textbf{10}) + (8 \times \underline{\hspace{1cm}})$

$8 \times \textbf{19} = 80 + \underline{\hspace{1cm}}$

$8 \times \textbf{19} = \underline{\hspace{1cm}}$

4. 4 × 36

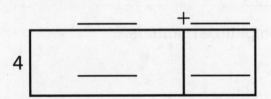

$4 \times \textbf{36} = (4 \times \underline{\hspace{1cm}}) + (4 \times \underline{\hspace{1cm}})$

$4 \times \textbf{36} = \underline{\hspace{1cm}} + \underline{\hspace{1cm}}$

$4 \times \textbf{36} = \underline{\hspace{1cm}}$

Use partial products to solve the problem.

5. The screen on Walker's cell phone is 7 centimeters by 14 centimeters. What is the area of the cell phone's screen?

_____ square centimeters

6. Lloyd's patio is 8 feet by 29 feet. What is the area of his patio? Draw an area model to show your thinking.

_____ square feet

Provide opportunities for your child to solve multiplication equations that involve multiplying a 2-digit by a 1-digit number. For example, have your child measure the length and width of a rectangular object. Then tell your child to create an area model to help him or her find the area of the object.

Additional Practice

Name _____

Review

You can use place value and properties of multiplication to multiply two multiples of 10. Place value can be used to explain patterns in the products of two multiples of 10.

A box contains 20 bags of popcorn. How many bags of popcorn are in 40 boxes? You can use 20×40 to find the total number of bags.

One way use basic facts and place value.	Another way use the Associative Property of Multiplication
$2 \times 40 = 2 \times 4$ tens	Use multiplication to decompose each multiple of 10. Then regroup the factors to solve the equation.
$2 \times 40 = 8$ tens	
$2 \times 40 = 80$	
20 is 10 times 2, so	
$20 \times 40 = 10 \times 80$	$20 \times 40 = 2 \times 10 \times 4 \times 10$
$ = 800$	$ = 2 \times 4 \times 10 \times 10$
	$ = 8 \times 10 \times 10$
	$ = 800$

How can you find the product? Complete the equation.

1. $60 \times 30 = 6 \times 10 \times$ _____ $\times 10$

$ = 6 \times$ _____ $\times 10 \times 10$

$ =$ _____ $\times 100$

$ =$ _____

2. $40 \times 80 = 40 \times$ _____ tens

$ =$ _____ tens

$ =$ _____

3. $70 \times 50 = 7 \times 10 \times$ _____ $\times 10$ **4.** $60 \times 80 = 60 \times$ _____ tens

$= 7 \times$ _____ $\times 10 \times 10$ $=$ _____ tens

$=$ _____ $\times 100$ $=$ _____

$=$ _____

5. A booklet of stamps contains 20 stamps. How many stamps are in 60 booklets?

6. A food bank collected 50 food items everyday for 30 days. How many total food items did they collect? Show and explain two ways to solve the problem.

7. Evie solves 50×80 using 5×8. She thinks there should be two zeroes in the product. Hudson thinks there should be three zeroes. How would you respond to Evie and Hudson?

Identify greater numbers around your home or in magazines and online articles. Ask your child to round the number to the nearest hundred thousand, ten thousand, and thousand. Then ask your child to identify which estimate is most reasonable in each situation.

Additional Practice

Name _____

Review

You can use multiplication strategies you know to help you solve multi-step word problems.

Lina is buying supplies to make bracelets. A package of beads costs $12 and a spool of wire costs $14. What is the total cost of 5 packages of beads and 3 spools of wire?

Use a variable to write an equation that represents the problem.

$(5 \times 12) + (3 \times 14)$

Step 1: Find the cost of beads

$5 \times 12 = 5 \times (10 + 2)$

$\qquad = (5 \times 10) + (5 \times 2)$

$\qquad = 50 + 10$

$\qquad = 60$

Step 2: Find the cost of wire

$3 \times 14 = 3 \times (10 + 4)$

$\qquad = (3 \times 10) + (3 \times 4)$

$\qquad = 30 + 12$

$\qquad = 42$

Step 3: Find the total cost

Add the costs to find the total cost. $60 + 42 = 102$

The total cost of 5 packages of beads and 3 spools of wire is $102

---

1. The cost of admission to the natural history museum for adults is $16. The cost of admission for a child is $11. What is the total cost of admission for 4 adults and 6 children?

2. A daycare orders 20 boxes of veggies straws and 18 boxes of crackers. Each box of veggies straws contains 12 bags and each boxes of crackers contains 16 bags. How many total bags of snacks did they order?

3. Omar has $150 to buy new netting for the soccer nets. He buys 9 yards of netting. Each yard of netting costs $14. How much money does he have left after buying the netting?

4. A school has 8 boxes of notebooks and 11 boxes of folders left after distributing them to all the students. There are 15 notebooks in each box and 18 folders in each box. How many notebooks and folders are left? Kelsey wrote an equation to solve. How would you respond to Kelsey? Explain and then solve.

$t = (8 \times 11) + (15 \times 18)$

Math @ Home Activity

Look up prices for the local movie theater. Ask your child to find the price of tickets for adults and children. Have students find the cost of 4 adults and 8 children attending a movie.

Additional Practice

Name _____

Review

You can use multiplication strategies you know to help you solve multi-step word problems.

Lina is buying supplies to make bracelets. A package of beads costs $12 and a spool of wire costs $14. What is the total cost of 5 packages of beads and 3 spools of wire?

Use a variable to write an equation that represents the problem.

$c = (12 \times 5) + (3 \times 14)$

Step 1: Find the cost of beads

$5 \times 12 = 5 \times (10 + 2)$

$\quad\quad\quad = (5 \times 10) + (5 \times 2)$

$\quad\quad\quad = 50 + 10$

$\quad\quad\quad = 60$

Step 2: Find the cost of wire

$3 \times 14 = 3 \times (10 + 4)$

$\quad\quad\quad = (3 \times 10) + (3 \times 4)$

$\quad\quad\quad = 30 + 12$

$\quad\quad\quad = 42$

Step 3: Find the total cost

Add the costs to find the total cost. $60 + 42 = 102$

The total cost of 5 packages of beads and 3 spools of wire is $102

---

1. The cost of admission to the natural history museum for adults is $16. The cost of admission for a child is $11. What is the total cost of admission for 4 adults and 6 children?

2. A daycare orders 20 boxes of veggies straws and 18 boxes of crackers. Each box of veggies straws contains 12 bags and each boxes of crackers contains 16 bags. How many total bags of snacks did they order?

3. Omar has $150 to buy new netting for the soccer nets. He buys 9 yards of netting. Each yard of netting costs $14. How much money does he have left after buying the netting?

4. A school has 8 boxes of notebooks and 11 boxes of folders left after distributing them to all the students. There are 15 notebooks in each box and 18 folders in each box. How many notebooks and folders are left? Kelsey wrote an equation to solve. How would you respond to Kelsey? Explain and then solve.

$t = (8 \times 11) + (15 \times 18)$

Math @ Home Activity

Look up prices for the local movie theater. Ask your child to find the price of tickets for adults and children. Have students find the cost of 4 adults and 8 children attending a movie.

Additional Practice

Name _____

Review

You can use partial quotients to solve division problems with four-digit dividends. Begin by distributing groups of 10, 100, and 1,000.

To find $7,092 \div 6$, you can give out 6 groups of 1,000, 6 groups of 100, 6 groups of 80, and 6 groups of 2 to find that the quotient is 1,182.

```
6 ) 7,092
   -6,000   1,000
    1,092
   -  600     100
      492
   -  480      80
       12
   -   12       2
        0   1,182
```

Use partial quotients to find the unknown.

1. $2,180 \div 5 = y$

     $y =$ _____

2. $4,936 \div 4 = z$

     $z =$ _____

3. $5,376 \div 3 = x$

     $x =$ _____

4. $7,389 \div 9 = w$

     $w =$ _____

Solve the word problem. Then, complete the sentence.

5. Ansel read 7 books and a total of 1,785 pages over the summer. If each book has the same number of pages, how many pages does each book have?

   Each book has _____ pages.

6. There were 2,225 tickets that sold for a play. The same number of people attended each of 5 performances. How many people attended each performance?

   The number of people that attended each performance was _____.

7. Shalonda has 4,650 pictures on 6 memory cards. The same number of pictures is on each memory card. How many pictures are on each memory card?

   There are _____ pictures on each memory card.

8. Terry claims 8,216 ÷ 8 has a 3-digit quotient. Is Terry correct? Explain your reasoning.

Math @ Home Activity

Write division equations, like the ones of the first page, on small pieces of paper. Use 8-10 pieces of paper. Place the pieces of paper in a basket or bag. Have your child pick on piece of paper from the container. Have him or her solve the equation before picking another piece of paper from the container.

Additional Practice

Name _____

Review

You can have remainders, or leftovers, when you divide.

To find $352 \div 6$, you can give out 6 groups of 50, 6 groups of 5, 6 groups of 2, and 6 groups of 1 to find that the quotient is 58 with a remainder of 4.

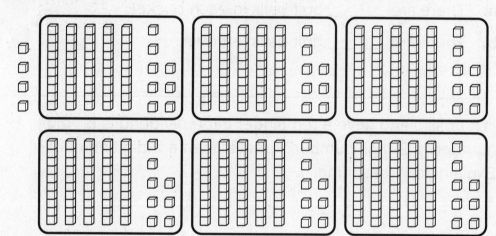

So, $352 \div 6 = 58$ R4.

Use partial quotients to find the unknown.

1. $451 \div 7 = p$

   $p =$ _____

2. $839 \div 5 = s$

   $s =$ _____

3. $2{,}999 \div 9 = q$

   $q =$ _____

4. $3{,}736 \div 3 = r$

   $r =$ _____

Solve the word problem. Then, complete the sentences.

5. Ava has saved $112 to buy ribbon for crafts. Each spool of ribbon costs $5. How many spools of ribbon can Ava buy?

   a. Ava can buy _____ spools of ribbon.

   b. How much money will Ava have left over? _____

6. A golf course uses 2,608 golf balls for its driving range. When all the golf balls are collected, they are split evenly into 9 buckets. How many golf balls are in each bucket?

   a. There are _____ golf balls in each bucket.

   b. There are _____ golf balls left over.

7. Is it possible to have a remainder if the dividend is a 1-digit number? a 2-digit number? greater than a 2-digit number? Explain your reasoning.

Ask your child for reasons why the length of a room would be divided into equal parts (for example, to place wall decorations at equal distances from each other). Then measure the length of a room in your home in inches (round to the nearest inch, if needed). Tell your child the room's length. Then ask your child to perform calculations with paper and pencil to divide the room's length into 3 equal parts, 4 equal parts, etc., up to 9 equal parts. Have your child calculate the remainder in each case. Then ask your child to check his or her answers by placing objects on the floor to represent the divisions.

Additional Practice

Name _____

Review

You can use the context of a problem to interpret the remainder of a multi-digit division problem.

Angie is placing 86 photographs in a photo album. Each page holds 3 photographs. To find how many pages she needs, divide 86 by 3.

$$86 \div 3 = 28 \text{ R2}$$

$$
\begin{array}{r}
3\overline{)86} \\
-60 \\
\hline
26 \\
-24 \\
\hline
2
\end{array}
\quad
\begin{array}{r}
20 \\
8 \\
\hline
28
\end{array}
$$

Angie will fill 28 pages of the photo album with photographs and will have 2 photographs left over. Since she needs to put the leftover photographs on a page, she will need one more page. Therefore, Angie will need 29 pages for 86 photographs.

---

1. William has 37 model cars to display equally on 5 shelves.

   a. How many model cars will be displayed on each shelf?

   Each shelf will display _____ model cars.

   b. Will William have any model cars left over? How do you know?

Solve the word problem. Then, complete the sentence.

2. Mark is using a wheelbarrow to move 123 bags of sand. Mark will place 8 bags of sand in the wheelbarrow for each load. How many loads will it take for Mark to move all the bags of sand?

It will take _____ loads to move all the bags of sand.

3. Shima wants to distribute 1,400 raffle tickets evenly among 6 classrooms. How many raffle tickets will each classroom get?

Each classroom will get _____ raffle tickets.

4. Xiomara says that after she places her 47 seashells equally in 4 glass jars, she will not have any seashells left over. How do you respond to Xiomara?

Math @ Home Activity

Fill a container with 76 fluid ounces (9$\frac{1}{2}$ cups) of water. Ask your child how many scoops with an 8-ounce (1 cup) measuring cup it will take to empty the container. Have your child use paper and pencil to set up and solve a division problem (76 ÷ 8 =?) to answer the question. Have your child explain what the remainder means. Then have your child use the measuring cup to confirm the answer to the division problem.

Additional Practice

Name _____

Review

You can use multiple steps to solve problems. You can use the problem situation to decide how to interpret remainders.

Maria is planting flowers in pots. She puts 4 flowers in each pot. She had 85 flowers. She gave 18 flowers to her neighbor. How many pots can she plant now?

Use a bar diagram to help find the number of flowers she gave to her neighbor.

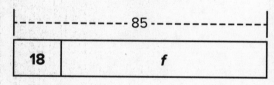

$f = 85 - 18$

$f = 67$

Maria has 67 flowers for the pots.

Then, divide to find the number of pots Maria can plant.

$p = 67 \div 4$

$$
\begin{array}{r}
4\overline{)67} \\
-60 \\
\hline
7 \\
-4 \\
\hline
3
\end{array}
\quad
\begin{array}{r}
15 \\
\\
1 \\
\hline
16
\end{array}
$$

$p = 16 \text{ R}3$

Maria can pot 16 pots with 4 flowers each. The remainder, 3, represents how many flowers are left over.

Solve. Show your work.

1. A company buys 125 new chairs for their sales team and 200 new chairs for other employees in their office building. If each floor of the building gets 25 chairs, how many floors are in the building?

2. Cooper has 112 pictures in one photo album and 162 pictures in another photo album. There are 12 pictures on each page. How many pages has he filled?

3. A school district orders new notebooks for fourth and fifth graders. They order 213 notebooks for fourth graders and 187 notebooks for fifth graders. If each classroom gets 20 notebooks, how many classrooms are there?

4. There are 164 fourth graders and 178 fifth graders going on a field trip. Each bus can hold 45 students. How many buses do they need for the field trip? Will all the buses be full?

Identify greater numbers around your home or in magazines and online articles. Ask your child to round the number to the nearest hundred thousand, ten thousand, and thousand. Then ask your child to identify which estimate is most reasonable in each situation.

Additional Practice

Name _____

Review

You can partition the whole into smaller pieces to represent an equivalent fraction. This changes the number of parts in the fraction and the whole.

Riley eats $\frac{1}{2}$ of an apple.

Everett eats $\frac{2}{4}$ of an apple.

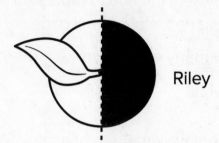

Riley

These fractions are equivalent because they are the same amount of the same size whole.

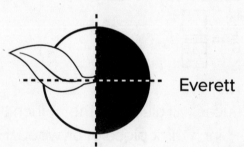

Everett

---

1. Shade to represent each fraction. Circle the two fractions that are equivalent.

$\frac{1}{3}$

$\frac{2}{6}$

$\frac{3}{8}$

2. Draw a fraction model or number line to justify that $\frac{3}{5}$ and $\frac{6}{10}$ are a pair of equivalent fractions.

3. Sadie measures a piece of fabric that is $\frac{2}{3}$ yard long. Write an equivalent fraction to represent the length of fabric if each third is partitioned into 4 equal parts.

$$\frac{2}{3} = \frac{\square}{\square}$$

4. Dwight walked $\frac{2}{8}$ lap of a track. Write an equivalent fraction to represent the distance that Dwight walked if each 2 eighths are combined into fourths.

$$\frac{2}{8} = \frac{\square}{\square}$$

5. Lucas will glue together $\frac{1}{8}$ inch thick pieces of plywood to make a $\frac{3}{4}$ inch thick piece of plywood. How many $\frac{1}{8}$ inch thick pieces should he glue together? Use pictures, words, or numbers to explain your reasoning.

Math @ Home Activity

Have your child use different sizes of measuring cups to identify equivalent fractions. For example, have your child determine if $\frac{1}{2}$ cup is equivalent to $\frac{4}{8}$ cup by pouring one $\frac{1}{2}$-cup of water in one container and four $\frac{1}{8}$-cups into a second identical container. If the water levels are the same, the fractions are equivalent.

Copyright © McGraw-Hill Education

Additional Practice

Name _____

Review

You can use the benchmark numbers $\frac{1}{2}$ and 1 to compare fractions. Compare $\frac{2}{6}$ and $\frac{6}{10}$ using the benchmark number $\frac{1}{2}$.

You know that $\frac{3}{6}$ is equal to $\frac{1}{2}$, and $\frac{5}{10}$ is equal to $\frac{1}{2}$.

$\frac{2}{6}$ is less than $\frac{1}{2}$, and $\frac{6}{10}$ is greater than $\frac{1}{2}$. So, $\frac{2}{6} < \frac{6}{10}$.

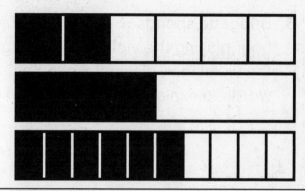

Write *less than* or *greater than* to compare each fraction to $\frac{1}{2}$. Then write $>$ or $<$ to compare the fractions.

1. $\frac{3}{4}$ is _____ $\frac{1}{2}$.

$\frac{3}{10}$ is _____ $\frac{1}{2}$.

$\frac{3}{4}$ ◯ $\frac{3}{10}$

2. $\frac{2}{5}$ is _____ $\frac{1}{2}$.

$\frac{7}{12}$ is _____ $\frac{1}{2}$.

$\frac{2}{5}$ ◯ $\frac{7}{12}$

Write *less than* or *greater than* to compare each fraction to 1. Then write > or < to compare the fractions.

3. $\frac{9}{10}$ is _____ 1.

   $\frac{6}{5}$ is _____ 1.

   $\frac{9}{10}$ ◯ $\frac{6}{5}$

4. $\frac{15}{12}$ is _____ 1.

   $\frac{6}{8}$ is _____ 1.

   $\frac{15}{12}$ ◯ $\frac{6}{8}$

---

5. Bridgette spends $\frac{4}{12}$ of an hour playing basketball and $\frac{5}{8}$ of an hour playing the violin. Does Bridgette spend more time playing the violin than playing basketball? Use pictures, numbers, or words to explain your reasoning.

6. Philip runs $\frac{11}{10}$ lap and Howard runs $\frac{5}{6}$ lap around a track. Who runs farther? Use pictures, numbers, or words to explain your reasoning.

7. Write one fraction that is greater than 1 and one fraction that is less than 1. Compare the fractions using pictures, numbers, or words to explain your reasoning.

Math @ Home Activity

Help your child practice comparing fractions to benchmark numbers by considering fractions around your home. First, identify fractions around your home. Then have your child determine whether each fraction is less than or greater than $\frac{1}{2}$ and whether each fraction is less than or greater than 1. Fractions around your home might be that $\frac{1}{3}$ of your utensils are spoons, $\frac{5}{6}$ of the rooms are painted tan, or $\frac{3}{4}$ of the family members are females.

Additional Practice

Name _____

Review

You can directly compare fractions that have like numerators or like denominators. To compare two fractions that do not have like numerators or like denominators, rewrite one or both of the fractions so that the equivalent fractions have the same numerator or the same denominator.

Compare $\frac{5}{8}$ and $\frac{10}{12}$ by finding an equivalent fraction with a like numerator.

$$\frac{10}{12} \text{ is equivalent to } \frac{5}{6}.$$

Eighths are smaller than sixths, so $\frac{5}{8}$ is less than $\frac{5}{6}$.

Use like numerators or like denominators to compare the fractions. Write > or < to compare the fractions.

1. $\frac{2}{5}$ ◯ $\frac{3}{5}$

2. $\frac{7}{6}$ ◯ $\frac{4}{6}$

3. $\frac{2}{10}$ ◯ $\frac{2}{8}$

4. $\frac{3}{4}$ ◯ $\frac{3}{10}$

5. Compare $\frac{1}{4}$ and $\frac{4}{12}$ using a like denominator. Write $\frac{1}{4}$ as an equivalent fraction with the same denominator as $\frac{4}{12}$. Then compare the fractions.

$\dfrac{\boxed{}}{12}$ ◯ $\dfrac{4}{12}$

Compare the original fractions.

$\dfrac{1}{4}$ ◯ $\dfrac{4}{12}$

6. Jaime and Vanessa are bowling. During the last game, $\frac{3}{10}$ of Jaime's balls were in the gutter, and $\frac{1}{5}$ of Vanessa's balls were in the gutter. Who had more balls in the gutter? Show your work and explain your reasoning.

7. Andrew and Carol are mowing lawns of the same size. Andrew has mowed $\frac{3}{5}$ of his yard, and Carol has mowed $\frac{4}{6}$ of her yard. Who has mowed more of his or her yard? Show your work and explain your reasoning.

Help your child practice comparing fractions that have like numerators or like denominators by helping him or her partition food to represent the fractions. Then your child can compare the amounts that represent each fraction. For example, ask your child to compare $\frac{2}{4}$ and $\frac{2}{3}$ by cutting slices of bread to represent each fraction. Tell your child to compare the amounts to determine which fraction is greater.

Additional Practice

Name _____

Review

You can add fractions with like denominators by adding the numerators and keeping the denominator the same.

You add the numerators because the number of parts that represent each fraction are being combined. You keep the denominator the same because the size of the parts has not changed.

$$\frac{2}{5} + \frac{1}{5} = \frac{3}{5}$$

Complete each equation.

1. $\frac{2}{4} + \frac{1}{4} = \dfrac{\boxed{}}{\boxed{}}$

2. $\frac{3}{6} + \frac{1}{6} = \dfrac{\boxed{}}{\boxed{}}$

3. $\frac{4}{8} + \frac{5}{8} = \dfrac{\boxed{}}{\boxed{}}$

4. $\frac{1}{3} + \frac{2}{3} + \frac{2}{3} = \dfrac{\boxed{}}{\boxed{}}$

5. $\frac{2}{10} + \frac{4}{10} + \frac{3}{10} = \dfrac{\boxed{}}{\boxed{}}$

6. $\frac{5}{12} + \frac{1}{12} + \frac{6}{12} = \dfrac{\boxed{}}{\boxed{}}$

7. Which three fractions have a sum of $\frac{13}{6}$?

$\frac{3}{6}$ $\frac{4}{6}$ $\frac{10}{6}$ $\frac{7}{6}$ $\frac{6}{6}$ $\frac{1}{6}$

8. Maxwell mixes $\frac{2}{8}$ gallon of white paint with $\frac{3}{8}$ gallon of black paint to make gray paint. How much gray paint does Maxwell have?

9. Blanca jogs $\frac{5}{10}$ mile in the morning, $\frac{4}{10}$ mile at noon, and $\frac{8}{10}$ mile in the evening. How far does Blanca jog each day? Explain.

10. Dennis uses $\frac{3}{12}$ yard of blue fabric, $\frac{2}{12}$ yard of red fabric, and $\frac{2}{12}$ yard of yellow fabric to make a small banner. He says he uses more than $\frac{1}{2}$ yard of fabric in all. Is Dennis correct? Explain your reasoning.

Math @ Home Activity

Have your child write the fractions $\frac{1}{12}, \frac{2}{12}, \frac{3}{12}, ..., \frac{12}{12}$ on index cards and place them facedown in an array. Then have your child try to find matches by identifying pairs of fractions that have a sum of $\frac{13}{12}$. Your child can also try to find matches with other sums or use other denominators.

Additional Practice

Name _____

Review

You can represent the subtraction of fractions with like denominators by separating parts that refer to the same whole.

$$\frac{5}{6} - \frac{3}{6} = \frac{2}{6}$$

The same subtraction equation can be represented on a number line.

$$\frac{5}{6} - \frac{3}{6} = \frac{2}{6}$$

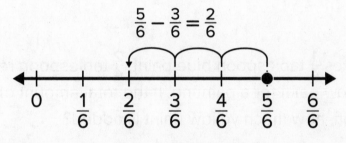

Use a representation to find each difference.

1. $\frac{5}{5} - \frac{3}{5} = \dfrac{\boxed{}}{\boxed{}}$

2. $\frac{7}{8} - \frac{4}{8} = \dfrac{\boxed{}}{\boxed{}}$

3. Use the number line to find the difference.

$\frac{7}{10} - \frac{5}{10} = \dfrac{\boxed{}}{\boxed{}}$

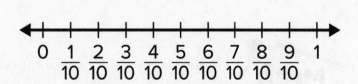

4. Lamar cuts $\frac{2}{12}$ yard off a board that is $\frac{9}{12}$ yard long. How long is the board now? Draw a picture to show your thinking.

5. Vanessa had $\frac{7}{8}$ of a tank of gas in her car before taking a trip. After the trip, she had $\frac{5}{8}$ of a tank of gas in the car. How much gas did Vanessa use on the trip? Explain.

6. Derek mixes $\frac{1}{4}$ tablespoon blue paint, $\frac{2}{4}$ tablespoon red paint, and some yellow paint for a painting. If the total amount of paint is $\frac{5}{4}$ tablespoon, how much yellow paint is added?

Math @ Home Activity

Have your child use a standard egg carton to practice fraction subtraction. Ask your child to place objects in a number of spaces to represent a fraction with a denominator of 12. Then have your child subtract a fraction by removing objects from some of the spaces.

Additional Practice

Name _____

Review

You can subtract fractions with like denominators by subtracting the numerators and keeping the denominator the same.

You subtract the numerators because the numerator counts the number of parts. You keep the denominator the same because the size of the parts has not changed.

$$\frac{7}{8} - \frac{2}{8} = \frac{5}{8}$$

Complete each equation.

1. $\dfrac{4}{3} - \dfrac{1}{3} = \dfrac{\square}{\square}$

2. $\dfrac{4}{5} - \dfrac{\square}{\square} = \dfrac{2}{5}$

3. $\dfrac{5}{6} - \dfrac{3}{6} = \dfrac{\square}{\square}$

4. $\dfrac{7}{10} - \dfrac{\square}{\square} = \dfrac{4}{10}$

5. $\dfrac{6}{8} - \dfrac{2}{8} - \dfrac{1}{8} = \dfrac{\square}{\square}$

6. $\dfrac{11}{12} - \dfrac{\square}{\square} - \dfrac{1}{12} = \dfrac{5}{12}$

7. Which two fractions have a difference of $\dfrac{3}{12}$?

   $\dfrac{6}{12}$ $\dfrac{8}{12}$ $\dfrac{2}{12}$ $\dfrac{13}{12}$ $\dfrac{1}{12}$ $\dfrac{9}{12}$

8. Chad purchased $\frac{4}{8}$ gallon of paint for a project. When he finished his project there was $\frac{1}{8}$ gallon of paint left. How much paint did Chad use for the project?

9. Sean used $\frac{2}{5}$ bag of cement mix to repair his steps and the rest of the bag to repair the sidewalk. How much cement mix did Sean use to repair the sidewalk? Explain.

10. Yeral used a total of $\frac{18}{6}$ cup of tomato paste while cooking. He used $\frac{4}{6}$ cup on pizza, $\frac{8}{6}$ on spaghetti, and the rest on ravioli. How much tomato paste did he use on the ravioli? Explain your reasoning.

Math @ Home Activity

Write out the fractions $\frac{1}{12}$ to $\frac{12}{12}$ on twelve equal-sized pieces of paper or index cards, one fraction on each. Have your child practice subtracting fractions by identifying pairs of fractions with a difference of $\frac{6}{12}$.

Additional Practice

Name _____

Review

You can solve problems involving the addition and subtraction of fractions by using equations to represent the given information.

Erin uses $\frac{7}{10}$ load of mulch in her front yard. She uses $\frac{4}{10}$ load less in her backyard than in her front yard. How much mulch does Erin use in her backyard?

You can use the information given in the problem to write a subtraction or an addition equation to represent the problem.

$$\frac{7}{10} - \frac{4}{10} = ?$$
$$\frac{7}{10} - \frac{4}{10} = \frac{3}{10}$$

$$? + \frac{4}{10} = \frac{7}{10}$$
$$\frac{3}{10} + \frac{4}{10} = \frac{7}{10}$$

The unknown quantity in each equation is $\frac{3}{10}$. So Erin uses $\frac{3}{10}$ load of mulch in her backyard.

Determine the number described.

1. $\frac{3}{5}$ less than $\frac{8}{5}$

2. $\frac{2}{6}$ greater than $\frac{3}{6}$

3. $\frac{7}{8}$ less than the sum of $\frac{2}{8}$ and $\frac{9}{8}$

4. $\frac{4}{12}$ more than the difference of $\frac{6}{12}$ and $\frac{3}{12}$

5. Mark read $\frac{2}{12}$ of a book on Monday and $\frac{3}{12}$ more of the book on Tuesday. He read the rest of his book on Wednesday. How much of the book did Mark read on Wednesday?

6. Rachael is making a bead necklace. In Rachael's necklace, $\frac{3}{8}$ of the beads are red and $\frac{2}{8}$ of the beads are orange. Rachael plans on using yellow beads to finish making her necklace. What fraction of the beads in Rachael's necklace will be yellow?

7. There is $\frac{5}{6}$ tank of gas in Dwight's car. He uses $\frac{1}{6}$ tank to go shopping and $\frac{2}{6}$ tank to visit a friend. Then Dwight drives home. There is $\frac{1}{6}$ tank in the car after Dwight drives home. How much gas did Dwight use to drive home? Explain.

8. Erica drank $\frac{2}{10}$ gallon more water on Friday than she drank on Thursday. Erica drank $\frac{5}{10}$ gallon of water on Thursday. How much water did Erica drink on Thursday and Friday combined? Explain your reasoning.

Math @ Home Activity

Use measuring cups or spoons and ask your child to find a fractional amount more or less than the given size. For example, ask your child to determine the amount that is $\frac{1}{4}$ cup less than $\frac{3}{4}$ cup.

Additional Practice

Name _____

Review

You can add mixed numbers by combining the whole-number parts and the fraction parts in any order.

To find $2\frac{3}{5} + 2\frac{3}{5}$, add the whole-number parts and the fraction parts.

$$2\frac{3}{5} + 2\frac{3}{5} = 4\frac{6}{5}$$

You can make another whole with the sum of the fraction parts and write an equivalent mixed number as the sum.

$$4\frac{6}{5} = 5\frac{1}{5}$$

You can represent the sum using fraction tiles or a number line.

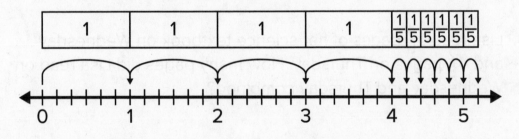

Find each sum. Use pictures, words, or numbers to show your work.

1. $1\frac{2}{6} + 1\frac{4}{6}$

2. $2\frac{3}{12} + 1\frac{1}{12}$

Find each sum. Use pictures, words, or numbers to show your work.

3. $2\frac{2}{3} + 2\frac{2}{3}$

4. $4\frac{3}{5} + 2\frac{4}{5}$

5. $3\frac{2}{10} + 2\frac{9}{10}$

6. $1\frac{11}{12} + 2\frac{5}{12}$

Solve each problem. Use pictures, words, or numbers to show your work.

7. Anthony uses $2\frac{3}{6}$ cups white flour and $1\frac{5}{6}$ cups wheat flour in a recipe. How much flour does Anthony use in all?

8. Lisa read $3\frac{4}{5}$ pages of her science textbook on Wednesday and $4\frac{2}{5}$ pages on Thursday. How many pages did Lisa read on Wednesday and Thursday combined?

9. Joel hiked a path that is $3\frac{5}{8}$ miles long on Saturday. He hiked a path that is $5\frac{7}{8}$ miles long on Sunday. Joel says he hiked more than 10 miles on Saturday and Sunday. Is Joel correct? Explain your reasoning.

Math @ Home Activity

Use a ruler or tape measure to find the lengths of objects around your home with your child. Find the lengths as mixed numbers to the nearest eighth of an inch. Then have your child find the combined length of different pairs of objects.

Additional Practice

Name _____

Review

To add mixed numbers, you can decompose each mixed number into whole-number parts and fraction parts or you can find an equivalent fraction for each mixed number and add the fractions.

You can find $1\frac{2}{3} + 3\frac{2}{5}$ in different ways.

$1\frac{2}{3} + 3\frac{2}{3}$

$1 + \frac{2}{3} + 3 + \frac{2}{3}$

$1 + 3 + \frac{2}{3} + \frac{2}{3}$

$4 + \frac{4}{3}$

$4 + \frac{3}{3} + \frac{1}{3}$

$4 + 1 + \frac{1}{3} = 5\frac{1}{3}$

$1\frac{2}{3} + 3\frac{2}{3}$

$\frac{5}{3} + \frac{11}{3} = \frac{16}{3}$

$\frac{16}{3} = \frac{3}{3} + \frac{3}{3} + \frac{3}{3} + \frac{3}{3} + \frac{3}{3} + \frac{1}{3}$

$\frac{16}{3} = 1 + 1 + 1 + 1 + 1 + \frac{1}{3} = 5\frac{1}{3}$

Find each sum. Use pictures, words, or numbers to show your work.

1. $2\frac{2}{5} + 2\frac{4}{5}$

2. $4\frac{5}{6} + 1\frac{3}{6}$

3. $2\frac{7}{10} + 3\frac{5}{10}$

4. $2\frac{7}{12} + 1\frac{2}{12}$

Solve each problem. Use pictures, words, or numbers to show your work. Check your answers for reasonableness.

5. Isabelle practices playing her guitar $2\frac{3}{4}$ hours one week $1\frac{2}{4}$ and hours the next week. How many hours does Isabelle practice playing her guitar during the two weeks?

6. Olivia buys $2\frac{3}{8}$ pounds of peanuts. James buys $1\frac{6}{8}$ pounds more peanuts than Olivia. How many pounds of peanuts does James buy?

7. Leeland paints $2\frac{5}{12}$ rooms on Monday and $3\frac{3}{12}$ rooms on Tuesday. How many rooms does Leeland paint on Monday and Tuesday combined?

8. Jashay's work to solve $3\frac{4}{5} + 4\frac{3}{5}$ is shown. Explain any errors you see in his work. Find the correct solution.

$$3\frac{4}{5} + 4\frac{3}{5} = \frac{3}{5} + \frac{4}{5} + \frac{4}{5} + \frac{3}{5}$$
$$= \frac{14}{5}$$
$$= \frac{10}{5} + \frac{4}{5}$$
$$= 2\frac{4}{5}$$

Math @ Home Activity

Help your child create a visual representation for the method they prefer to use to add mixed numbers. Keep the representation as simple and visual as possible.

Additional Practice

Name _____

Review

You can solve word problems involving addition and subtraction of mixed numbers by using representations and equations to find the solutions.

Bruce ran $1\frac{1}{4}$ laps around a track. Then he ran another $1\frac{2}{4}$ laps. What is the total number of laps Bruce ran?

You can represent the problem on a number line. Start at $1\frac{1}{4}$ and count forward $1\frac{2}{4}$, or $\frac{6}{4}$.

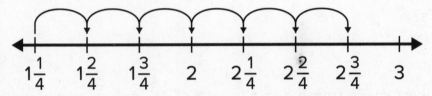

You can also write an equation to represent the problem.

First Run		Second Run		Total
$1\frac{1}{4}$	$+$	$1\frac{2}{4}$	$=$	?
$1\frac{1}{4}$	$+$	$1\frac{2}{4}$	$=$	$2\frac{3}{4}$

Bruce ran $2\frac{3}{4}$ laps in all.

---

1. Drew spends $3\frac{2}{3}$ minutes solving one math problem and $4\frac{2}{3}$ minutes solving another math problem. How many minutes does Drew spend solving both math problems? Use pictures, words, or numbers to show your work.

Solve each problem. Use pictures, words, or numbers to show your work.

2. Jessica uses $2\frac{4}{5}$ bags of mulch around her mailbox and some in her flower beds. She uses a total of $6\frac{2}{5}$ bags of mulch. How many bags of mulch does Jessica use in her flower beds?

3. Erin's hair is 9 inches long. After a haircut, her hair is $7\frac{7}{10}$ inches long. How much of the length was trimmed off?

4. Cara's paper airplane flew $9\frac{7}{12}$ feet. Andy's paper airplane flew $6\frac{11}{12}$ feet. How much farther did Cara's paper airplane fly than Andy's?

5. A rectangular picture frame has a length of $4\frac{3}{8}$ inches and a width of $5\frac{3}{8}$ inches. What is the perimeter of the picture frame? Explain your reasoning.

Math @ Home Activity

Work with your child to make a chart that shows the different strategies that can be used to solve word problems with addition and subtraction of mixed numbers.

Additional Practice

Name _____

Review

You can use repeated addition to multiply a unit fraction by a whole number.

Farah uses $\frac{1}{3}$ sheet of stickers to decorate each page of her scrapbook.

If she decorates 6 scrapbook pages, how many sheets of stickers does she use?

$6 \times \frac{1}{3}$ is the same as the sum of 6 groups of $\frac{1}{3}$.

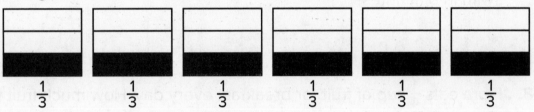

$$\frac{1}{3} + \frac{1}{3} + \frac{1}{3} + \frac{1}{3} + \frac{1}{3} + \frac{1}{3} = \frac{6}{3} \text{ or } 2$$

Since $\frac{1}{3} + \frac{1}{3} + \frac{1}{3} + \frac{1}{3} + \frac{1}{3} + \frac{1}{3}$ and $6 \times \frac{1}{3}$ both represent the same amount, $6 \times \frac{1}{3} = \frac{6}{3}$ or 2 sheets. Farah needs 2 sheets of stickers.

Rewrite each multiplication equation as repeated addition.

1. $5 \times \frac{1}{6} =$ _____

_____ + _____ + _____ + _____ + _____ = _____

2. $7 \times \frac{1}{5} =$ _____

_____ + _____ + _____ + _____ + _____ + _____ + _____

= _____

Solve.

3. $4 \times \dfrac{1}{2} =$ _____ or 2

4. _____ $= 6 \times \dfrac{1}{4}$

5. $9 \times \dfrac{1}{8} =$ _____

6. _____ $= 8 \times \dfrac{1}{5}$

---

7. Monroe swims $\dfrac{1}{3}$ lap in one minute. How many laps will Monroe swim in 9 minutes?

8. Flora eats $\dfrac{1}{4}$ cup of fruit for breakfast every day. How much fruit will she eat in 5 days?

9. Maryrose is creating an octagon for a room decoration. The sides of the octagon will be created with pieces of ribbon. Each side of the octagon is $\dfrac{1}{2}$ foot long. If she has 5 feet of ribbon, will she have enough ribbon to make the decoration? Explain.

Math @ Home Activity

Help your child identify unit fractions around your home. Common unit fractions can be found in recipes and distance measurements. Create word problems that require your child to multiply the identified unit fraction by a whole number. Ask him or her to write the multiplication expression as repeated addition before solving the problem.

Additional Practice

Name _____

Review

You can use different representations to show multiplying a fraction by a whole number.

Use a number line to find $4 \times \frac{2}{3}$.

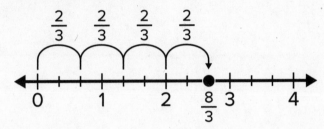

The number line shows 4 jumps of $\frac{2}{3}$. The last jump lands on $\frac{8}{3}$.

$4 \times \frac{2}{3} = \frac{8}{3}$

Use number lines to solve.

1. $3 \times \frac{4}{5} =$ _____

2. $5 \times \frac{1}{2} =$ _____

3. $7 \times \frac{6}{10} =$ _____

Create a fraction model to solve.

4. $9 \times \frac{1}{4} = $ _____

5. $7 \times \frac{3}{6} = $ _____

6. $6 \times \frac{7}{8} = $ _____

7. Each plant on a porch needs $\frac{3}{4}$ gallon of water each day. If there are 5 plants on the porch, how much water is needed each day?

8. It takes Evan $\frac{1}{5}$ hour to read each chapter in his book. There are 9 chapters in the book. Evan claims it will take him 2 hours to read the entire book. How would you respond to Evan?

Math @ Home Activity

While cooking dinner, have your child identify the fractional amounts of ingredients that will be needed. Ask your child to determine the amount of each ingredient that will be needed if the recipe is doubled or tripled. Have him or her use number lines or fraction models to show his or her thinking.

Additional Practice

Name _____

Review

You can use what you know about multiplication and multiplying a unit fraction by a whole number to multiply a fraction by a whole number.

$$6 \times \frac{2}{3} = ?$$

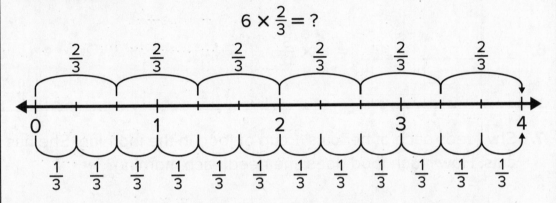

You can show 6 jumps of $\frac{2}{3}$ on a number line to represent the problem.

You can write $\frac{2}{3}$ as a multiple of a unit fraction. Each jump of $\frac{2}{3}$ can also be shown as 2 jumps of $\frac{1}{3}$ or $2 \times \frac{1}{3}$.

$6 \times \frac{2}{3}$ is equal to $6 \times 2 \times \frac{1}{3}$.

$$6 \times 2 \times \frac{1}{3} = ?$$
$$12 \times \frac{1}{3} = \frac{12}{3} \text{ or } 4$$

Decompose the fraction into unit fractions. Then solve.

1. $5 \times \frac{3}{5} =$ _____

_____ × _____ × _____ = _____

2. $6 \times \frac{5}{8} =$ _____

_____ × _____ × _____ = _____

Solve.

3. $3 \times \dfrac{3}{4} = $ _____

4. _____ $= 7 \times \dfrac{5}{6}$

5. $4 \times \dfrac{7}{10} = $ _____

6. _____ $= 8 \times \dfrac{5}{12}$

---

7. Shy feeds each of her cats $\dfrac{3}{4}$ cup of food in the morning. She has 3 cats. How much food does she need each morning?

8. A carpenter uses $\dfrac{9}{10}$ gallon of stain for every 3 picnic tables he makes. How much stain will he need for 12 picnic tables?

9. Sunny is finding the product of 8 and $\dfrac{7}{8}$. He rewrites $8 \times \dfrac{7}{8}$ as $8 \times \dfrac{1}{7} \times \dfrac{1}{8}$. How do you respond to Sunny?

Math @ Home Activity

Create situations in which your child can practice multiplying fractions by whole numbers. For example, if each person in your family drinks $\dfrac{2}{3}$ cup of milk for breakfast, have your child determine how much milk is needed for the entire family.

Additional Practice

Name _____

Review

You can use representations and equations to find the answer to a word problem that involves multiplying a fraction or mixed number by a whole number.

A recipe for granola calls for $\frac{7}{8}$ cup of nuts. If the recipe is made 4 times, how many cups of nuts will be needed?

$$4 \times \frac{7}{8} = ?$$

Show 4 groups of $\frac{7}{8}$ on a number line.

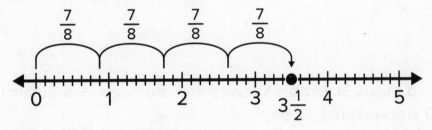

$3\frac{1}{2}$ cups of nuts will be needed.

Use the number line to solve.

1. Each of 5 apples weighs $\frac{3}{5}$ pound. What is the total weight of the apples?

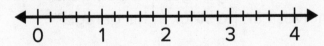

2. Lisa swims $2\frac{1}{3}$ miles each day. How many miles will Lisa swim in 4 days?

3. Randa uses $4\frac{1}{2}$ gallons of water each time she washes dishes. If she washes dishes 5 times, will she use less than 20 gallons of water? If she uses more than 20 gallons, how much more will she use?

4. Jordan wants to drink 10 cups of water each day. She drinks $2\frac{3}{4}$ cups at each of 3 meals. How many more cups of water does she need to drink to reach 10 cups?

5. Conrad builds stackable shelves that are each $1\frac{5}{6}$ feet tall. How tall will 3 stacked shelves be?

He wants to stack the shelves in his garage. The garage is 12 feet tall. How many shelves can he stack in his garage?

Math @ Home Activity

Create a word problem about an everyday situation that requires multiplying a fraction or mixed number by a whole number. Ask your child to show you how to draw a number line to solve the word problem. Ask questions that require your child to correct your thinking.

Additional Practice

Name _____

Review

You can write fractions with denominators of 10 or 100 as a decimal by using a decimal point to separate the whole number part from the fractional part.

Write a fraction and a decimal for the representation.

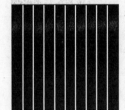 

$\frac{11}{10}$ or $1\frac{1}{10}$

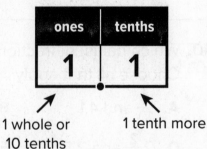

1 whole or 10 tenths

1 tenth more

The fraction $\frac{11}{10}$ or $1\frac{1}{10}$ is the same as the decimal 1.1.

Write a fraction and decimal for the representation.

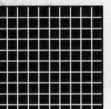 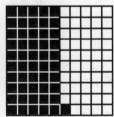

$1\frac{51}{100}$

ones	tenths	hundredths
1	5	1

1 whole or 100 hundredths

50 hundredths or 5 tenths

1 hundredth more

The fraction $1\frac{51}{100}$ can also be written as the decimal 1.51.

Write a fraction and a decimal for each representation.

1.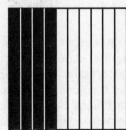

2.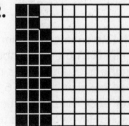

3. ▮▯▯▮▮▮▮▮▮▯   ▮▮▮▮▮▮▮▯▮▯   ▮▮▯▯▯▯▯▯

Write each amount as a decimal.

4. $2\frac{79}{100}$

5. six hundredths

6. one and 2 tenths

7. $1\frac{5}{10}$

8. $2\frac{9}{10}$

9. $1\frac{7}{100}$

---

10. Which pair(s) of fractions and decimals are **not** equivalent? Choose all that apply.

  A. $\frac{4}{10}$ and 4.1   **B.** 2.6 and $26\frac{3}{10}$ **C.** 3.7 and $3\frac{7}{10}$

  D. $2\frac{2}{10}$ and 2.2   **E.** $1\frac{141}{100}$ and 1.41   **F.** 1.79 and $1\frac{79}{100}$

11. Winter sells 3.5 pounds of apples to a customer. Write the weight as a fraction.

12. A pumpkin weighs $1\frac{60}{100}$ pounds. Jae says the pumpkin weighs 1.6 pounds. How do you respond to Jae?

Create a matching game for your child to play. Write 10 decimal numbers on self-stick notes, 1 number per note. Each decimal number should have a digit in the ones and tenths places or ones, tenths and hundredths. For example, you could write 0.8, 1.6, 3.15, etc. Write the equivalent fractions on separate self-stick notes, 1 fraction per note. Arrange the notes, blank side up, on a table. Have your child turn two notes over. If they are equivalent, have them stick the notes together. If they are not equivalent, have them turn them back over and turn over 2 different notes. Have your child repeat until all matches are found.

Additional Practice

Name _____

Review

You can solve word problems by converting larger units of measure to smaller units.

Guy's bucket holds 10 gallons of water. There are 10 quarts of water in the bucket already. How many more quarts of water should be added to fill the bucket?

```
|--------- 10 gallons or 40 quarts ---------|
| 10 quarts |         ? quarts              |
```

$40 - 10 = 30$ quarts

So, 30 quarts of water should be added to fill the bucket.

Solve.

1. Lydia's puppy has a mass of 8 kilograms. Her cat has a mass of 3,850 grams. What is the difference in mass of Lydia's dog and cat in grams?

   _____ grams

2. A play starts at 6:30 p.m. and ends at 8:25 p.m. How many minutes long is the play?

   _____ minutes

3. Jody needs to knit 13 yards of yarn. She knits 9 feet of yarn each day for 4 days. How many more feet of yarn does she need to knit?

   _____ feet

Solve.

4. Gina's water bottle holds 2 liters of water. She drinks 750 milliliters of water at basketball practice. How many milliliters of water are left in her water bottle?

_____ milliliters

5. Elena runs $3\frac{1}{2}$ miles on Sunday and 5 miles on Monday. How many more feet does Elena run on Monday than on Sunday?

_____ feet

6. Liam's mom gives him 5 quarts of apple juice. Liam needs 13 pints of apple juice. How many more pints of apple juice does he need?

_____ pints

7. Rosemary lives 8 kilometers from the library. Callie lives 6.5 kilometers from the library. How many meters closer is Callie's house than Rosemary's house to the library?

_____ meters

_____

8. Lloyd has $10.75 to buy 3 gallons of milk. The store sells quarts of milk for $1. How much more money does Lloyd need? Explain.

Math @ Home Activity

While grocery shopping, use the signs around the store to create word problems that require your child to convert between measurement units. For example, if you need gallons of juice and the juice is only sold in quarts, ask your child to determine how many quarts you will need.

Additional Practice

Name _____

Review

You can use representations to solve word problems that involve converting units of measure.

Elliot practices the piano for $1\frac{1}{2}$ hour each week and keeps track of how long he practices each day in his log. How many more minutes does Elliot need to practice this week?

Multiply to convert a larger unit to a smaller unit.

Convert hours to minutes

weekly minutes $= 1\frac{1}{2} \times 60 = 90$ minutes

$90 - 60 = 30$ minutes

Find the total number of minutes practiced so far.

$T = 22 + 20 + 18 = 60$

Elliot has practiced 60 minutes so far this week.

Elliot needs to practice for 30 more minutes this week.

Day of the week	Minutes Practiced
Monday	22
Tuesday	20
Wednesday	18

What is the solution to the problem?

1. Ellie makes 4 gallons of apple sauce. She puts the apple sauce in 1-pint containers. How many containers can she fill?

2. Lisa's hockey practice lasts 1 hour and 31 minutes. Practice ends at 5:46 p.m. What time did practice start?

3. Hannah ran 810 yards in 4 minutes. Justin ran 1,800 feet in the same amount of time. Who ran faster?

4. Henry buys a 3-pound bag of flour. Eliza already has $2\frac{1}{2}$ pounds of flour. How many ounces of flour do they have together?

5. Heather buys $5\frac{3}{4}$ yards of green fabric for the school play costumes. She uses 30 inches of green fabric for each costume. How many costumes can she make?

6. A tomato plant grows $\frac{3}{4}$ of a foot each month. How many inches does it grow in 4 months?

Identify objects of capacity in your home such as a milk or juice carton. Ask your child to convert it to a smaller or larger capacity. Next measure an object or space in your home and ask your child to convert to a larger or smaller unit.

Lesson 13-9
Additional Practice

Name _____

> **You can solve real-world problems involving area and perimeter.**
>
> A park needs mulch to cover a playground. What is the area of the playground the mulch will need to cover?
>
>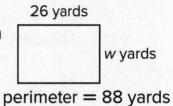
> 26 yards
> *w* yards
> perimeter = 88 yards
>
> To find the area of the playground, you need to find the width. Substitute what you know into a formula for perimeter. Recall that the sum of two adjacent sides of a rectangle is half of the perimeter.
>
> $l + w = \frac{1}{2} \times P$
>
> $l + w = \frac{1}{2} \times 88$
>
> $l + w = 44$
>
> $26 + w = 44$
>
> $26 + 18 = 44 \rightarrow w = 18$
>
> The area of the playground is $A = l \times w = 26 \times 18 = 468$ square yards.

1. The perimeter of the pool is 52 yards. What is the area?

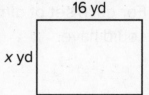

16 yd

x yd

2. Find the perimeter of the garden.

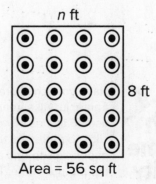

n ft

8 ft

Area = 56 sq ft

3. Goldie is building a butterfly sanctuary. She has 28 feet of mesh she can use for the perimeter of the sanctuary.

   a. What whole-number dimensions can she use to build the sanctuary?

   b. What set of dimensions would give the largest area for the sanctuary?

   c. What do you notice about the dimensions that give the butterflies the largest area?

4. Lonnie has 32 inches of wood to make a wooden rectangular picture frame.

   a. What are all of the possible whole-number dimensions for the picture frame?

   b. For each set of dimensions, find the area the picture frame could have.

Have your child pretend he or she is going to build a rectangular picture frame, just like Lonnie does in Exercise 4. Tell your child that he or she will use 26 inches of wood to make the picture frame. Give your child 26 paper clips to represent the 26 inches of wood. Have your child manipulate the paper clips to find all of the possible whole-number dimensions for his or her picture frame. Tell your child to write all of the dimensions on paper. Then, for each set of dimensions, tell your child to find the area of the picture frame using the formula $A = l \times w$.

Additional Practice

Name _____

You can display measurement data in a line plot.

The table shows the lengths of fish. Represent this data in a line plot.

Fish Lengths (in.)

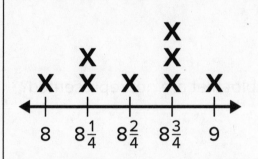

Fish Lengths (in.)	
$8\frac{3}{4}$	8
$8\frac{1}{4}$	$8\frac{3}{4}$
9	$8\frac{1}{4}$
$8\frac{2}{4}$	$8\frac{3}{4}$

---

1. The table shows the lengths of Cameron's color pencils. Represent this data in a line plot.

Color Pencil Lengths (in.)

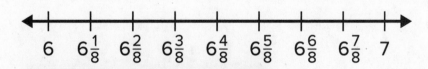

Color Pencil Lengths (in.)	
$6\frac{3}{8}$	$6\frac{4}{8}$
$6\frac{5}{8}$	$6\frac{1}{8}$
$6\frac{7}{8}$	7
6	$6\frac{5}{8}$
$6\frac{5}{8}$	$6\frac{2}{8}$

Use your line plot to answer the following questions.

2. Which length was the least common? How many color pencils are this long?

    a. _____ inches

    b. ___ color pencils

3. Which length was the most common? How many color pencils are this long?

    a. _____ inches

    b. ___ color pencils

4. Are there any lengths on your line plot that are not represented? Explain.

---

5. Shari is deciding whether to collect measurement data to the nearest eighth inch or quarter inch to create a line plot. Shari wants her line plot to be as accurate as possible. Which measurement data should Shari collect? Explain.

Have your child measure the lengths of eight leaves to the nearest eighth of an inch. Have him or her create a line plot using the measurements he or she collected. Ask him or her questions about the data.

Additional Practice

Name _____

You can use data in a line plot to solve problems involving addition and subtraction of fractions.

The line plot shows the height of 9 sunflowers in a garden. What is the difference in height between the tallest and shortest sunflowers?

Sunflower Height (ft)

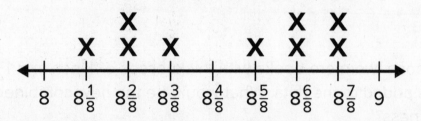

To find the difference in height, subtract the shortest height from the tallest height. Then use addition to check your answer.

$$8\frac{7}{8} - 8\frac{1}{8} = \frac{6}{8}$$

$$\frac{6}{8} + 8\frac{1}{8} = 8\frac{7}{8}$$

The difference in height between the tallest and shortest sunflowers is $\frac{6}{8}$ foot.

---

1. The line plot shows the thickness of books on a shelf. What is the difference between the thicknesses of the two thinnest books?

Book Thickness (in.)

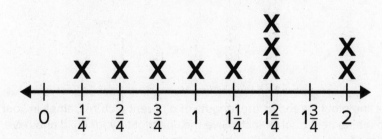

_____ inch

2. What is the difference between the thicknesses of the thickest book and the thinnest book?

_____ inches

3. What is the combined thickness of all the books?

_____ inches

4. Suppose two more books with thicknesses of 1 inch and $1\frac{3}{4}$ inches were added to the data. What would be the new combined thickness?

_____ inches

5. A line plot shows one X above $\frac{1}{4}$, three Xs above $\frac{2}{4}$, and one X above $\frac{3}{4}$ to show how much time it took Caleb to complete his math homework one week. Caleb says that he spent a total of $2\frac{1}{4}$ hours working on math homework that week. Is Caleb correct? Explain.

Math @ Home Activity

Create a line plot that shows the lengths of different picture frames in your home to the nearest quarter inch. Give the line plot to your child and have him or her answer questions like the following.

What is the difference in length between the longest and shortest picture frames?

Additional Practice

Name _____

Review

You can draw geometric figures, such as lines, line segments, and rays.

Draw line *AB*.

Draw line segment *CD*.

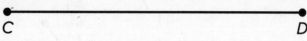

Draw ray *EF*.

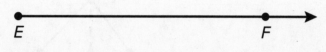

Draw each figure.

1. ray *ST*   **2.** segment *JK*   **3.** line *QR*

Name each figure.

4.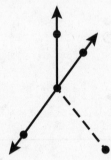

_____

5.

_____

Identify the dashed part of each figure as a line, line segment, or ray.

6.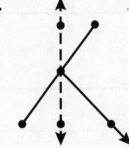

_____

7.

_____

8. Draw and label a figure that contains line *NO*, ray *OP*, and segment *NP*.

Math @ Home Activity

Give your child food items to represent line segments and points. For example, you can give your child pretzel sticks to represent line segments and grapes to represent points. Ask your child to use the food items to construct lines, line segments, and rays. Ask them to justify each figure they create.

Additional Practice

Name _____

Review

You can classify an angle as acute if it has a measure that is less than the measure of a right angle, obtuse if it has a measure that is greater than the measure of a right angle, and right if the amount of rotation from one ray of the angle to the other is $\frac{1}{4}$ of a whole circle.

Classify each angle.

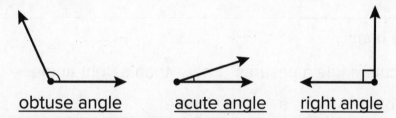

obtuse angle   acute angle   right angle

Draw a line from each figure to its matching description.

1.

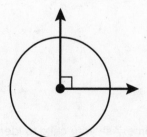

rotation is more than $\frac{1}{4}$ of a whole circle

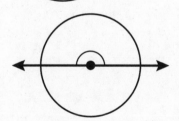

rotation is $\frac{1}{4}$ of a whole circle

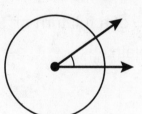

rotation is less than $\frac{1}{4}$ of a whole circle

Identify each angle as acute, obtuse, or right.

2.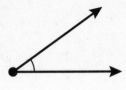

_____

3.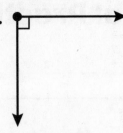

_____

4.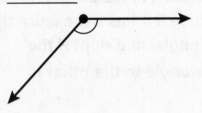

_____

5.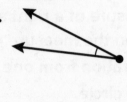

_____

Fill in the blank.

6. An acute angle measures _____ than a right angle.

7. A $\frac{1}{4}$ rotation forms a _____ angle.

8. A _____ angle measures greater than a right angle.

9. Caroline draws an angle whose measure is $\frac{1}{3}$ of a full turn of a circle. How can she classify the angle? Explain.

Math @ Home Activity

Have your child identify angles inside your home. Ask your child to identify each angle as obtuse, right, or acute. For example, the angle formed by two adjacent sides of a window is a right angle.

Additional Practice

Name _____

Review

You can use a protractor to measure and to draw angles.

To measure an angle with a protractor, first place the center of the protractor on the endpoint of the angle.

Next, line up one ray with 0° on the protractor.

Then find the tick mark on the protractor that aligns with the second ray of the angle.

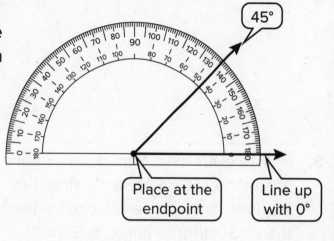

45°

Place at the endpoint

Line up with 0°

Use a protractor to measure each angle.

1.

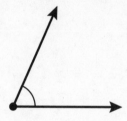

_____

2.

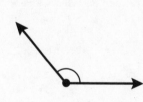

_____

3.

_____

4.

_____

Draw an angle for each measure.

5. 78°

6. 29°

7. 90°

8. 162°

9. Before exercising, Sam does a sequence of stretches. One of Sam's stretches requires him to raise his legs to greater than a 30 degree angle. Is Sam performing the stretch properly? Explain.

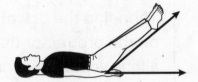

Math @ Home Activity

Have your child use benchmark angles to estimate the measures of angles he or she sees in the pictures of a book or magazine. Then have your child find each actual angle measure and state whether each angle is obtuse, acute, or right.

Additional Practice

Name _____

Review

You can use lines to describe the relationship between paths. Lines can be classified by the way they intersect.

Zoe is riding her bike along trails. How can you describe the relationship between the paths?

Partridge Trail and Quail Trail are parallel because they never meet or intersect and are always the same distance apart.

Dove Trail is perpendicular to Partridge Trail and Quail Trail. Perpendicular lines intersect to form a right angle.

How can you describe the pair of lines shown? Label each pair as parallel, perpendicular or neither.

1.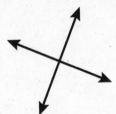

    _____

2.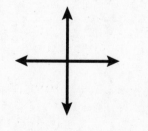

    _____

3.

    _____

4.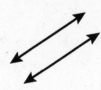

    _____

Draw a pair of lines that match the description.

5. Parallel

6. Perpendicular

7. Intersecting, but not perpendicular

8. Wade thinks that a window only has perpendicular line segments. Explain any errors in his thinking.

Have your child identify different objects or furniture around the house that contains parallel or perpendicular lines. For example, the shelves of a bookcase are parallel lines.

Additional Practice

Name _____

Review

You can use a protractor to solve problems involving angles. An angle can be broken into smaller angles. The sum of the smaller angles is equal to the original angle.

You can use a protractor to find the measurement of the angle.

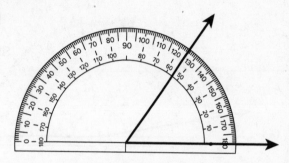

You can draw a ray inside the angle to break it into two smaller angles.

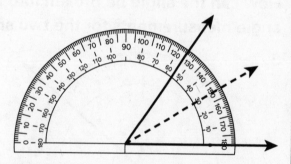

The measure of the first angle is 30 degrees. The measure of the second angle is 25 degrees. The sum of the angles is 55 degrees.

What is the sum of the two angles?

1.

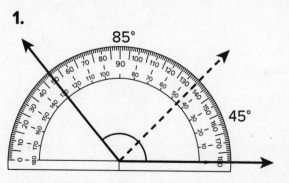
85°
45°

_____

2.

15°
20°

_____

What is the measure of the unknown angle?

3.

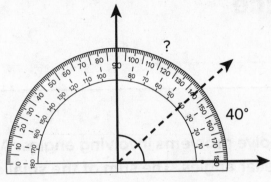

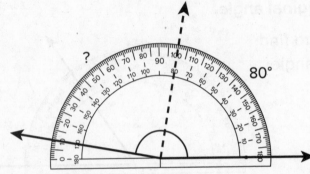

40°

4.

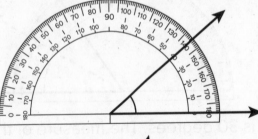

80°

How can the angle be broken into two smaller angles? Write possible angle measurements for the two smaller angles.

5.

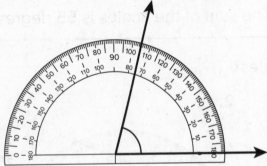

_____

6.

_____

Math @ Home Activity

Draw angles for your child to measure with a protractor. Then, once they have found the measurement ask them to split it into two smaller angles.

Lesson **14-6**

Additional Practice

Name _____

<table><tr><td>

</td></tr></table>

Review

You can write an addition or subtraction equation to calculate a missing angle measure.

The combined angle measure is 60°. Find x.

$40° + x = 60°$ $60° - 40° = x$

$\qquad x = 20°$ $\qquad 20° = x$

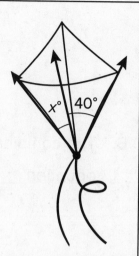

Find each combined angle measure.

1.

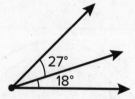

_____

2.

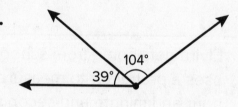

_____

3.

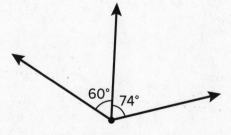

_____

4.

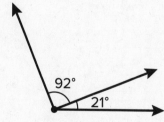

_____

Write an equation to find each unknown angle measure. Then solve the equation.

5. The combined angle measure is 94°.

equation: _____

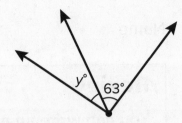

6. The combined angle measure is 142°.

equation: _____

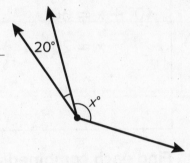

7. Elvira gets home from school at 4:17. She uses a protractor to measure the angles the hour and minute and second hands make. The angle the hour and seconds hand make is 102°. Find the value of x.

$x =$ _____

Create riddles about finding unknown angle measures for your child to solve by writing an addition or subtraction equation. For example, "The combined angle measure is 102°. One of the angle measures is 89°. What is the angle measure of the second angle?" Challenge them to write both an addition and subtraction equation for each riddle. Your child may find it helpful to draw a picture.

Lesson **14-7**

Additional Practice

Name _____

Review

You can classify quadrilaterals based on the absence or presence of parallel or perpendicular lines.

Some quadrilaterals have neither parallel nor perpendicular lines.

A quadrilateral with exactly one pair of parallel lines is a **trapezoid**.

A quadrilateral with two pairs of parallel lines is a **parallelogram**.

A parallelogram with 4 equal sides is a **rhombus**.

A parallelogram with 4 right angles and 2 pairs of equal opposite sides is a **rectangle**.

A parallelogram with 4 right angles and 4 equal sides is a **square**.

---

1. Match the quadrilateral with its attributes.

square	A quadrilateral with two pairs of parallel lines
rhombus	A parallelogram with 4 right angles and 2 pairs of equal opposite sides
rectangle	A quadrilateral with exactly one pair of parallel lines
parallelogram	A parallelogram with 4 right angles and 4 equal sides
trapezoid	A parallelogram with 4 equal sides

2. Which statements are true about all trapezoids?
Choose all that apply.

    A. All trapezoids have 4 sides.

    B. All trapezoids have 2 pairs of parallel lines.

    C. All trapezoids have 4 right angles.

    D. All trapezoids have 4 equal sides.

3. Which statements are true about all rhombuses?
Choose all that apply.

    A. All rhombuses have 4 sides.

    B. All rhombuses have 2 pairs of parallel lines.

    C. All rhombuses have 4 right angles.

    D. All rhombuses have 4 equal sides.

---

4. How are a square and a rectangle the same?
How are they different?

Have your child find two-dimensional figures around your home. Tell your child to identify whether each two-dimensional figure has parallel lines, perpendicular lines, or both parallel and perpendicular lines. Ask them to explain how they determined the types of lines the figures have.

Additional Practice

Name _____

Review

You can classify triangles by their angles and side lengths.

Classify the triangle by its angles and side lengths.

The triangle has one right angle. The triangle has no equal side lengths. The triangle is a right scalene triangle.

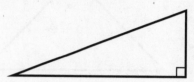

Classify each triangle by its angles. Write acute, right, or obtuse.

1.

_____

2.

_____

3.

_____

Classify each triangle by its side lengths. Write equilateral, scalene, or isosceles.

4.
5.
6.

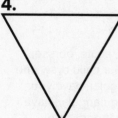 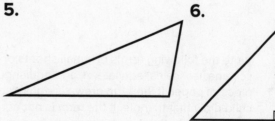

_____   _____   _____

Classify each triangle by its angles and side lengths.

7.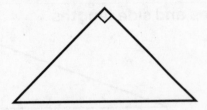

_____

8.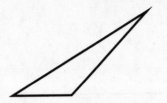

_____

9.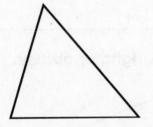

_____

10. A section of a park is shaped like the triangle below. If the section is a right isosceles triangle, it will be used for a baseball field. If it is an obtuse isosceles triangle, it will be used for a dog park. What will the section of the park be used for? Explain.

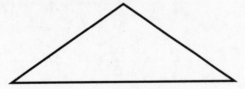

Write the following terms on small sheets of paper: right, acute, obtuse, scalene, isosceles, equilateral, and equiangular. Have your child draw one piece of paper. If the term drawn is equilateral or equiangular, have your child draw that triangle. If the term is not equilateral or equiangular, have them draw another piece of paper. Have your child draw the triangle described by the two terms. If the second term is equilateral or equiangular, have them draw another piece of paper.

Additional Practice

Name ....

Review

A figure has line symmetry if it can be folded over a line so that its two halves match exactly. This line is called a line of symmetry. You can draw lines of symmetry.

Draw lines of symmetry on the figure below.

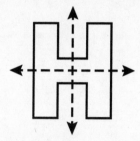

This figure has 2 lines of symmetry.

Determine if each dotted line shows a line of symmetry. Write *yes* or *no*.

1.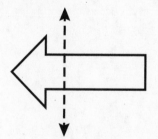

2.

3.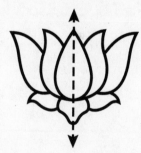

4.

5.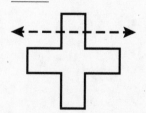

6.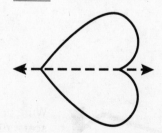

Draw the rest of each figure using its line of symmetry.

7.

8.

9. Leroy draws the pine tree in a mural. Use line symmetry to complete the tree.

Additional Practice

Name _____

Review

You can identify the lines of symmetry in a figure. To help, look for the attributes that are symmetrical.

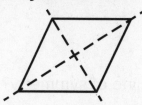

A rhombus has 2 pairs of parallel sides and all sides are the same length. These are symmetric attributes, so a rhombus has 2 lines of symmetry.

How many lines of symmetry are there? Draw the lines.

1.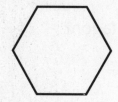

2.

3.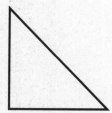

4.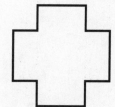

5. A triangle has three lines of symmetry. What kind of triangle could it be? Explain.

6. What are the attributes of a hexagon with 6 lines of symmetry that make it symmetrical. Select all that apply.

A. All 6 sides are equal

B. The opposite sides are parallel

C. The angles are equal

D. The sides are perpendicular

7. How many lines of symmetry does a circle have?

8. Do all triangles have a line of symmetry?

9. How many lines of symmetry are there in this pentagon?

Math @ Home Activity

Identify different objects around the house. Ask your child to identify any lines of symmetry within the object. Ask your child to explain how they found the lines of symmetry.